Problem-Solving Cases in Microsoft® Access and Excel, Third Annual Edition

JOSEPH A. BRADY,

ELLEN F. MONK

Australia • Canada • Mexico • Singapore • Spain • United Kingdom • United States

Problem-Solving Cases in Microsoft® Access and Excel, Third Annual Edition

Executive Editor:
Mac Mendelsohn

Senior Acquisitions Editor
Maureen Martin

Senior Product Manager:
Tricia Boyle

Development Editor:
DeVona Dors

Senior Marketing Manager:
Karen Seitz

Associate Product Manager:
Mirella Misiaszek

Editorial Assistant:
Jennifer Smith

Production Editors:
Bobbi Jo Frasca, Summer Hughes

Manufacturing Coordinator:
Laura Burns

Cover Designer:
Betsy Young

Cover Artist:
Rakefet Kenaan

Compositor:
Gex Publishing Services

Copyeditor:
Lori Cavanaugh

Copyeditor/Proofreader:
John Bosco

Indexer:
May Hasso

Printed in the United States of America

2 3 4 5 6 7 8 9 Glob 09 08 07 06 05

For more information, contact Course Technology, 25 Thomson Place, Boston, Massachusetts, 02210.

Or find us on the World Wide Web at: www.course.com

0-619-21697-2

DEDICATION

To our development editor, DeVona, who has made our books the best they can be

Preface

For the past 15 years, we have taught MIS courses at the University of Delaware. From the start, we wanted to use good computer-based case studies for the database and decision-support portions of our courses.

We could not find a casebook that met our needs! This surprised us because our requirements, we thought, were not unreasonable. First, we wanted cases that asked students to think about real-world business situations. Second, we wanted cases that provided students with hands-on experience, using the kind of software that they had learned to use in their computer literacy courses—and that they would later use in business. Third, we wanted cases that would strengthen students' ability to analyze a problem, examine alternative solutions, and implement a solution using software. Undeterred by the lack of casebooks, we wrote our own cases.

New! This book is the fourth casebook we have written for Course Technology. The cases are all new and the tutorials are updated. New features of this textbook reflect comments and suggestions that we have received from instructors who have used our casebook in their classrooms.

New! As with our prior casebooks, we include tutorials that prepare students for the cases, which are challenging but doable. Most of the cases are organized in a way that helps the student think about the logic of each case's business problem and then how to use the software to solve the business problem. Some instructors requested cases that require students to think entirely on their own to build a solution, and we have responded with two "Challenge" cases (Cases 6 and 13) that do that. The cases will fit well in an undergraduate MIS course, an MBA Information Systems course, or a Computer Science course devoted to business-oriented programming.

Book Organization

The book is organized into six parts:

1. Database Cases Using Access (also includes a "Challenge Case")
2. Decision Support Cases Using Excel Scenario Manager
3. Decision Support Cases Using the Excel Solver
4. Decision Support Cases Using Basic Excel Functionality (also includes a "Challenge Case")
5. Integration Case: Using Access and Excel
6. Presentation Skills

Part 1 begins with two tutorials that prepare students for the Access case studies. Parts 2 and 3 each begin with a tutorial that prepares students for the Excel case studies. All four tutorials provide students with hands-on practice in using the software's more advanced features—the kind of support that other books about Access and Excel do not give to the student. Part 4 asks students to use Excel's basic functionality for decision support. The Excel "Challenge Case" is also in this section—students must determine how far to go beyond basic functionality to reach a business solution. Part 5 challenges students to use both Access and Excel to find a solution to solve a business problem. Part 6 is a tutorial that hones students' skills in creating and delivering an oral presentation to business managers. The next section explores each of these parts in more depth.

Part 1: Database Cases

Using Access

This section begins with two tutorials and then presents six case studies.

Tutorial A: Database Design

This tutorial helps the student to understand how to set up tables to create a database, without requiring students to learn formal analysis and design methods, such as data normalization.

Tutorial B: Microsoft Access

The second tutorial teaches students the more advanced features of Access queries and reports—features that students will need to know to complete the cases.

Cases 1–6

Six database cases follow Tutorials A and B. The students' job is to implement each case's database in Access so form, query, switchboard, and report outputs can help management. The first case is an easier "warm-up" case. The next five cases require a more demanding database design and implementation effort.

In Cases 3 through 5, students are challenged to go beyond what is asked of them: They must analyze the business operation and then suggest (or implement) additional forms, queries, switchboards, and reports that will help the business operation to run more efficiently. By Case 6, the "Challenge Case," students are given very little guidance and must analyze the business problem and determine how to create all the desired outputs. Students must think more creatively and in far greater depth to solve this business problem.

Part 2: Decision Support Cases

Using the Excel Scenario Manager

This section has one tutorial and two decision support cases requiring the use of Excel Scenario Manager.

Tutorial C: Building a Decision Support System in Excel

This section begins with a tutorial using Excel for decision support and spreadsheet design. Fundamental spreadsheet design concepts are taught. Instruction on the Scenario Manager, which can be used to organize the output of many "what-if" scenarios, is emphasized.

Cases 7–8

These two cases can be done with or without the Scenario Manager (although the Scenario Manager is nicely suited to them). In each case, students must use Excel to model two or more solutions to a problem. Students then use the outputs of the model to identify and document the preferred solution via a memorandum and, if assigned to do so, an oral presentation.

Part 3: Decision Support Cases

Using Excel Solver

This section has one tutorial and two decision support cases requiring the use of Excel Solver.

Tutorial D: Building a Decision Support System Using Excel Solver

This section begins with a tutorial about using the Solver, which is a decision support tool for solving optimization problems.

Cases 9–10

Once again, in each case, students use Excel to analyze alternatives and identify and document the preferred solution.

Part 4: Decision Support Cases

Using Basic Excel Functionality

Cases 11–12

The cases continue with two cases that use basic Excel functionality, i.e., the cases do not require the Scenario Manager or the Solver. Excel is used to test the student's analytical skills in "what if" analyses.

Case 13: Challenge Case using Excel

This case requires the student to decide which modeling tools to use: Solver, Scenario Manager, or Excel's basic functionality alone. Working creatively, the student must design and implement a solution to the stated problem.

Part 5: Integration Case

Using Excel and Access

Case 14

This case integrates Access and Excel. This case is included because of a trend toward sharing data among multiple software packages to solve problems.

Part 6: Presentation Skills

Tutorial E: Giving an Oral Presentation

Because each case includes an assignment that gives students practice in making a presentation to management on the results of their analysis of the case, this section gives advice on how to create oral presentations. It also has technical information on charting and pivot tables, techniques that might be useful in case analyses or as support for presentations. This tutorial will help students to organize their recommendations, to present their solutions in both words and graphics, and to answer questions from the audience. For larger classes, instructors may wish to have students work in teams to create and deliver their presentations—which would model the "team" approach used by many corporations.

INDIVIDUAL CASE DESIGN

The format of the fourteen cases follows this template.

- Each case begins with a *Preview* of what the case is about and an overview of the tasks.
- The next section, *Preparation*, tells students what they need to do or know to complete the case successfully. (Of course, our tutorials prepare students for the cases!)
- The third section, *Background*, provides the business context that frames the case. The background of each case models situations that require the kinds of thinking and analysis that students will need in the business world.
- This is followed by the *Assignment* sections, which are generally organized in a way that helps students to develop their analyses. (The "Challenge" cases provide less guidance, however, than the other cases.)
- The last section, *Deliverables*, lists what the student must hand in: printouts, a memorandum, a presentation, and files on disk. The list is similar to the kind of deliverables that a business manager might demand.

Using the Cases

We have successfully used cases like these in our undergraduate MIS courses. We usually begin the semester with Access database instruction. We assign the Access database tutorials and then a case to each student. Then, for Excel DSS instruction, we do the same thing—assign a tutorial and then a case. Instructors with time for more casework could do that, but also follow up the Access cases with the Access "Challenge Case" and the Excel cases with the Excel "Challenge Case."

Microsoft Office 2003

New! Another important feature of the Third Annual Edition is the inclusion of a FREE 30-day trial of Microsoft Office 2003 in the back of every book. This CD-ROM contains the entire Microsoft Office 2003 suite.

Technical Information

This textbook was quality assurance tested using the Windows XP Professional operating system, Microsoft Access 2003, and Microsoft Excel 2003.

Data Files and Solution Files

We have created "starter" data files for the Excel cases, so students need not spend time typing in the spreadsheet skeleton. Case 14 also requires students to load a data file. All these files can be found on the Course Technology Web site, which is available to both students and instructors. Go to *www.course.com* and search for this textbook by title, author, or ISBN. You are granted a license to copy the data files to any computer or computer network used by individuals who have purchased this textbook.

Solutions to the material in the text are available to instructors. These can also be found at *www.course.com*. Search for this textbook by title, author, or ISBN. The solutions are password protected.

Instructor's Manual

An Instructor's Manual is available to accompany this text. The Instructor's Manual contains additional tools and information to help instructors successfully use this textbook. Items such as a Sample Syllabus, Teaching Tips, and Grading Guidelines are an example of the material that can be found in the Instructor's Manual. Instructors should go to *www.course.com* and search for this textbook by title, author, or ISBN. The Instructor's Manual is password protected.

Acknowledgements

We would like to give many thanks to the team at Course Technology, including to our Developmental Editor, DeVona Dors; Senior Product Manager, Tricia Boyle; and to our Production Editor, Bobbi Jo Frasca. As always, we acknowledge our students' diligent work.

Contents

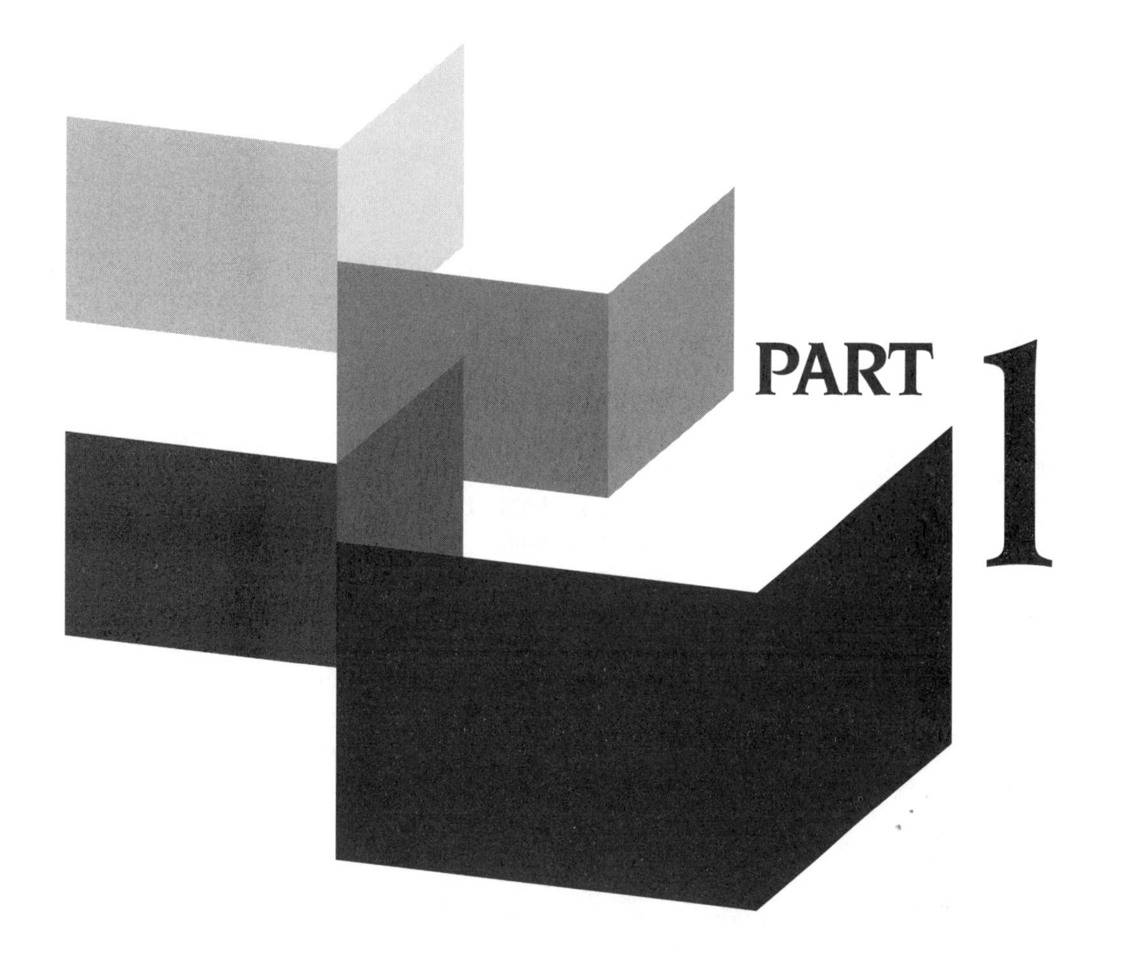

Database Cases Using Access

Database Design

This tutorial has three sections. The first section briefly reviews basic database terminology. The second section teaches database design. The third section has a practice database design problem.

Review of Terminology

Let's begin by reviewing some basic terms that will be used throughout this textbook. In Access, a **database** is a group of related objects that are saved into one file. An Access **object** can be a table, a form, a query, or a report. You can identify an Access database file because it has the suffix **.mdb**.

A **table** consists of data that is arrayed in rows and columns. A **row** of data is called a **record**. A **column** of data is called a **field**. Thus, a record is a set of related fields. The fields in a table should be related to one another in some way. For example, a company might have employee data in a table called EMPLOYEE. That table would contain data fields about employees—their names, addresses, etc. It would not have data fields about the company's customers—that data would go into a CUSTOMER table.

A field's values have a **data type**. When a table is defined, the nature of each field's data is declared. Then, when data is entered, the database software knows how to interpret each entry. Data types in Access include the following:

- "Text" for words
- "Integer" for whole numbers
- "Double" for numbers that can have a decimal value
- "Currency" for numbers that should be treated as dollars and cents
- "Yes/No" for variables that can have only two values (1-0, on/off, yes/no, true/false)
- "Date/Time" for variables that are dates or times

Each database table should have a **primary key** field, a field in which each record has a *unique* value. For example, in an EMPLOYEE table, a field called SSN (for Social Security Number) could be a primary key, because each record's SSN value would be different from every other record's SSN value. Sometimes, a table does not have a single field whose values are all different. In that case, two or more fields are combined into a **compound primary key**. The combination of the fields' values is unique.

Database tables should be logically related to one another. For example, suppose that a company has an EMPLOYEE table with fields for SSN, Name, Address, and Telephone Number. For payroll purposes, the company would also have an HOURS WORKED table with a field that summarizes Labor Hours for individual employees. The relationship between the EMPLOYEE table and the HOURS WORKED table needs to be established in the database; otherwise, how could you tell which employees worked which hours? This is done by including the primary key field from the EMPLOYEE table (SSN) as a field in the HOURS WORKED table. In the HOURS WORKED table, the SSN field is then called a **foreign key**.

Data can be entered into a table directly or by entering the data into a **form**, which is based on the table. The form then inserts the data into the table.

A **query** is a question that is posed about data in a table (or tables). For example, a manager might want to know the names of employees who have worked for the company more than five years. A query could be designed to interrogate the EMPLOYEE table in that way. The query would be "run" and its output would answer the question.

A query may need to pull data from more than one table, so queries can be designed to interrogate more than one table at a time. In that case, the tables must first be connected by a **join** operation, which links tables on the values in a field that they have in common. The common field acts as a kind of "hinge" for the joined tables; the query generator treats the joined tables as one large table when running the query.

In Access, queries that answer a question are called **select** queries. Queries can be designed that will change data in records or delete entire records from a table. These are called **update** and **delete** queries, respectively.

Access has a **report** generator that can be used to format a table's data or a query's output.

Database Design

"Designing" a database refers to the process of determining which tables need to be in the database and the fields that need to be in each table. This section begins with a discussion of design concepts. The following key concepts are defined:

- Entities
- Relationships
- Attributes

This section then discusses database design rules, a series of steps we advise that you use to build a database.

Database Design Concepts

Computer scientists have formal ways of documenting a database's logic, but learning the notations and mechanics can be quite time-consuming and difficult. Doing this usually takes a good portion of a Systems Analysis and Design course. This tutorial will teach you database design by emphasizing practical business knowledge. This approach will let you design serviceable databases. Your instructor may add some more formal techniques.

A database models the logic of an organization's operation, so your first task is to understand that operation. You do that by talking to managers and workers, by observation, and/or by looking at business documents, such as sales records. Your goal is to identify the business's "entities" (sometimes called *objects*, in yet another use of this term). An **entity** is some thing or some event that the database will contain. Every entity has characteristics, called **attributes**, and a **relationship(s)** to other entities. Let's take a closer look.

Entities

An entity is a tangible thing or event. The reason for identifying entities is that *an entity eventually becomes a table in the database*. Entities that are things are easy to identify. For example, consider a video store's database. The database would need to contain the names of videotapes and the names of customers who rent them, so you would have one entity named VIDEO and another named CUSTOMER.

By contrast, entities that are events can be more difficult to identify. This is probably because events cannot be seen, but they are no less real. In the video store example, one event would be the VIDEO RENTAL, and another would be HOURS WORKED by employees.

Your analysis is made easier by the knowledge that organizations usually have certain physical entities, such as:

- Employees
- Customers
- Inventory (Products)
- Suppliers

The database for most organizations would have a table for each of those entities. Your analysis is also made easier by the knowledge that organizations engage in transactions internally and with the outside world. These transactions are the subject of any accounting course, but most people can understand them from events in daily life. Consider the following examples:

- Organizations generate revenue from sales or interest earned. Revenue-generating transactions are event entities, called SALES, INTEREST, etc.
- Organizations incur expenses from paying hourly employees and purchasing materials from suppliers. HOURS WORKED and PURCHASES would be event entities in the databases of most organizations.

Thus, identifying entities is a matter of observing what happens in an organization. Your powers of observation are aided by knowing what entities exist in the databases of most organizations.

Relationships

The analyst should consider the relationship of each entity to other entities. For each entity, the analyst should ask, "What is the relationship, if any, of this entity to every other entity identified?" Relationships can be expressed in English. For example, a college's database might have entities for STUDENT (containing data about each student), COURSE (containing data about each course), and SECTION (containing data about each section). A relationship between STUDENT and SECTION would be expressed as "Students enroll in Sections."

An analyst must also consider what is called the **cardinality** of any relationship. Cardinality can be one-to-one, one-to-many, or many-to-many. These are summarized as follows:

- In a one-to-one relationship, one instance of the first entity is related to just one instance of the second entity.
- In a one-to-many relationship, one instance of the first entity is related to many instances of the second entity, but only one instance of the second entity is related to an instance of the first.

- In a many-to-many relationship, one instance of the first entity is related to many instances of the second entity, and one instance of the second entity is related to many of the first.

To make this more concrete, again think about the college database having STUDENT, COURSE, and SECTION entities. A course, such as Accounting 101, can have more than one section: 01, 02, 03, 04, etc. Thus:

- The relationship between the entities COURSE and SECTION is one-to-many. Each course has many sections, but each section is for just one course.
- The relationship between STUDENT and SECTION is many-to-many. Each student can be in more than one section because each student can take more than one course. Also, each section has more than one student.

Worrying about relationships and their cardinalities may seem tedious to you now. However, you will see that this knowledge will help you to determine the database tables needed (in the case of many-to-many relationships) and the fields that need to be shared between tables (in the case of one-to-many relationships).

Attributes

An attribute is a characteristic of an entity. You identify attributes of an entity because *attributes become a table's fields*. If an entity can be thought of as a noun, an attribute can be thought of as an adjective describing the noun. Continuing with the college database example, again think about the STUDENT entity. Students have names. Thus, Last Name would be an attribute, a field, of the STUDENT entity. First Name would be an attribute as well. The STUDENT entity would have an Address attribute, another field; and so on.

Sometimes, it is difficult to tell the difference between an attribute and an entity. One good way to differentiate them is to ask whether there can be more than one of the possible attribute for each entity. If more than one instance is possible, and you do not know in advance how many there will be, then it's an entity. For example, assume that a student could have two (but no more) Addresses—one for "home" and one for "on campus." You could specify attributes Address 1 and Address 2. On the other hand, what if the number of student addresses could not be stipulated in advance, but all addresses had to be recorded? You would not know how many fields to set aside in the STUDENT table for addresses. You would need a STUDENT ADDRESSES table, which could show any number of addresses for a student.

Database Design Rules

Your first task in database design is always to understand the logic of the business situation. You then build a database for the requirements of that situation. To create a context for learning about database design, let's first look at a hypothetical business operation and its database needs.

Example: The Talent Agency

Suppose that you have been asked to build a database for a talent agency. The agency books bands into nightclubs. The agent needs a database to keep track of the agency's transactions and to answer day-to-day questions. Many questions arise in running the business. For example, a club manager might want to know which bands are available on a certain date at a certain time or the agent's fee for a certain band. Similarly, the agent might want to see a list of all band members and the instrument each plays, or a list of all the bands having three members.

Suppose that you have talked to the agent and have observed the agency's business operation. You conclude that your database would need to reflect the following facts:

1. A "booking" is an event in which a certain band plays in a particular club on a particular date, starting at a certain time, ending at a certain time, and for a specific fee. A band can play more than once a day. The Heartbreakers, for example, could play at the East End Cafe in the afternoon and then at the West End Cafe that night. For each booking, the club pays the talent agent, who keeps a 5% fee and then gives the rest to the band.
2. Each band has at least two members and an unlimited maximum number of members. The agent notes a telephone number of just one band member, which is used as the band's contact number. No two bands have the same name or telephone number.
3. No band members in any of the bands have the same name. For example, if there is a Sally Smith in one band, there is no Sally Smith in any other band.
4. The agent keeps track of just one instrument that each band member plays. "Vocals" is an instrument for this record-keeping purpose.
5. Each band has a desired fee. For example, the Lightmetal band might want $700 per booking and would expect the agent to try to get at least that amount for the band.
6. Each nightclub has a name, an address, and a contact person. That person has a telephone number that the agent uses to contact the club. No two clubs have the same name, contact person name, or telephone number. Each club has a target fee. The contact person will try to get the agent to accept that amount for a band's appearance.
7. Some clubs will feed the band members for free, and others will not.

Before continuing, you might try to design the agency's database on your own. What are the entities? Recall that databases usually have CUSTOMER, EMPLOYEE, and INVENTORY entities and an entity for the revenue-generating transaction event. Each entity becomes a table in the database. What are the relationships between entities? For each entity, what are its attributes? These become the fields in each table. For each table, what is the primary key?

Six Database Design Rules

Assume that you have gathered information about the business situation in the talent agency example. Now you want to identify the tables for the database and then the fields in each table. To do that, observe the following six rules.

Rule 1: You do not need a table for the business itself. The database represents the entire business. Thus, in our example, Agent and Agency are not entities.

Rule 2: Identify the entities in the business description. Look for the things and events that the database must contain. These become tables in the database. Typically, certain entities are represented. In the talent agency example, you should be able to see these entities:

- *Things*: The product (inventory for sale) is Band. The customer is Club.
- *Events*: The revenue-generating transaction is Bookings.

You might ask yourself: Is there an EMPLOYEE entity? Also, isn't INSTRUMENT an entity? These issues will be discussed as the rules are explained.

Rule 3: Look for relationships between the entities. Look for one-to-many relationships between entities. The relationship between these entities must be established in tables, and this is done by using a foreign key. The mechanics of that is discussed in the next rule. (See the discussion of the relationship between Band and Band Member.)

Look for many-to-many relationships between entities. In each of these relationships, there is the need for a third entity that associates the two entities in the relationship. Recall the STUDENT—SECTION many-to-many relationship example. A third table is needed to show the ENROLLMENT of specific students in specific sections. The mechanics of doing this is discussed in the next rule. (See the discussion of the relationship between BAND and CLUB.) (Note that ENROLLMENT can also be thought of as an event entity, and you might have already identified this entity. Forcing yourself to think about many-to-many relationships means that you will not miss it.)

Rule 4: Look for attributes of each entity, and designate a primary key. Think of entities as nouns. List the adjectives of the nouns. These are the attributes which, as was previously mentioned, become the table's fields. After you have identified fields for each table, designate one as the primary key field, if one field has unique values. Designate a compound primary key if no one field has unique values.

The attributes, or fields, of the BAND entity are Band Name, Band Phone Number, and Desired Fee. No two band names can be the same, it is assumed, so the primary key field in this case can be Band Name. Figure A-1 shows the BAND table and its fields: Band Name, Band Phone Number, and Desired Fee; the data type of each field is also shown.

BAND	
Field	***Data Type***
Band Name (primary key)	Text
Band Phone Number	Text
Desired Fee	Currency

Figure A-1 The BAND table and its fields

Two BAND records are shown in Figure A-2.

Band Name (primary key)	*Band Phone Number*	*Desired Fee*
Heartbreakers	981 831 1765	$800
Lightmetal	981 831 2000	$700

Figure A-2 Records in the BAND table

If there could be two bands called the Heartbreakers in the agency, then Band Name would not be a good primary key. Some other unique identifier would be needed. Such situations are common in business. Most businesses have many types of inventory, and duplicate names are possible. The typical solution is to assign a number to each product to be used as the primary key field. For example, a college could have more than one faculty member with the same name, so each faculty member would be assigned a Personal Identification Number (PIN). Similarly, banks assign a PIN for each depositor. Each automobile that a car manufacturer makes gets a unique Vehicle Identification Number (VIN). Most businesses assign a number to each sale, called an invoice number. (The next time you buy something at a grocery store, note the number on your receipt. It will be different from the number that the next person in line sees on their receipt).

At this point, you might ask why Band Member would not be an attribute of BAND. The answer is that you must record each band member, but you do not know in advance how many members will be in each band. Therefore, you do not know how many fields to allocate to the BAND table for members. Another way to think about Band Member is that they are, in effect, the agency's employees. Databases for organizations usually have an EMPLOYEE entity. Therefore, you should create a BAND MEMBER table with the attributes Member Name, Band Name, Instrument, and Phone. The BAND MEMBER table and its fields are shown in Figure A-3.

BAND MEMBER	
Field Name	***Data Type***
Member Name (primary key)	Text
Band Name (foreign key)	Text
Instrument	Text
Phone	Text

Figure A-3 The BAND MEMBER table and its fields

Five records in the BAND MEMBER table are shown in Figure A-4.

Member Name (primary key)	***Band Name***	***Instrument***	***Phone***
Pete Goff	Heartbreakers	Guitar	981 444 1111
Joe Goff	Heartbreakers	Vocals	981 444 1234
Sue Smith	Heartbreakers	Keyboard	981 555 1199
Joe Jackson	Lightmetal	Sax	981 888 1654
Sue Hoopes	Lightmetal	Piano	981 888 1765

Figure A-4 Records in the BAND MEMBER table

Instrument can be included as a field in the BAND MEMBER table, because the agent records only one for each band member. Instrument can thus be thought of as a way to describe a band member, much as the phone number is part of the description. Member Name can be the primary key because of the (somewhat arbitrary) assumption that no two members in any band have the same name. Alternatively, Phone could be the primary key if it could be assumed that no two members share a telephone. Alternatively, a band member ID number could be assigned to each person in each band, which would create a unique identifier for each band member handled by the agency.

You might ask why Band Name is included in the BAND MEMBER table. The common sense reason is that you did not include the Member Name in the BAND table. You must relate bands and members somewhere, and this is the place to do it.

Another way to think about this involves the cardinality of the relationship between BAND and BAND MEMBER. It is a one-to-many relationship: One band has many members, but each member is in just one band. You establish this kind of relationship in the database by using the primary key field of one table as a foreign key in the other. In BAND MEMBER, the foreign key Band Name is used to establish the relationship between the member and his or her band.

The attributes of the entity CLUB are Club Name, Address, Contact Name, Club Phone Number, Preferred Fee, and Feed Band? The table called CLUB can define the CLUB entity, as shown in Figure A-5.

CLUB	
Field Name	***Data Type***
Club Name (primary key)	Text
Address	Text
Contact Name	Text
Club Phone Number	Text
Preferred Fee	Currency
Feed Band?	Yes/No

Figure A-5 The CLUB table and its fields

Two records in the CLUB table are shown in Figure A-6.

Club Name (primary key)	*Address*	*Contact Name*	*Club Phone Number*	*Preferred Fee*	*Feed Band?*
East End	1 Duce St.	Al Pots	981 444 8877	$600	Yes
West End	99 Duce St.	Val Dots	981 555 0011	$650	No

Figure A-6 Records in the CLUB table

You might wonder why Bands Booked Into Club (or some such field name) is not an attribute of the CLUB table. There are two answers. First, you do not know in advance how many bookings a club will have, so the value cannot be an attribute. Furthermore, BOOKINGS is the agency's revenue-generating transaction, an event entity, and you need a table for that business transaction. Let us consider the booking transaction next.

You know that the talent agent books a certain band into a certain club on a certain date, for a certain fee, starting at a certain time, and ending at a certain time. From that information, you can see that the attributes of the BOOKINGS entity are Band Name, Club Name, Date, Start Time, End Time, and Fee. The BOOKINGS table and its fields are shown in Figure A-7.

BOOKINGS	
Field Name	***Data Type***
Band Name	Text
Club Name	Text
Date	Date/Time
Start Time	Date/Time
End Time	Date/Time
Fee	Currency

Figure A-7 The BOOKINGS table and its fields—and no designation of a primary key

Some records in the BOOKINGS table are shown in Figure A-8.

Band Name	*Club Name*	*Date*	*Start Time*	*End Time*	*Fee*
Heartbreakers	East End	11/21/05	19:00	23:30	$800
Heartbreakers	East End	11/22/05	19:00	23:30	$750
Heartbreakers	West End	11/28/05	13:00	18:00	$500
Lightmetal	East End	11/21/05	13:00	18:00	$700
Lightmetal	West End	11/22/05	13:00	18:00	$750

Figure A-8 Records in the BOOKINGS table

No single field is guaranteed to have unique values, because each band would be booked many times, and each club would be used many times. Further, each date and time could appear more than once. Thus, no one field can be the primary key.

If a table does not have a single primary key field, you can make a compound primary key whose field values together will be unique. Because one band can be in only one place at a time, one possible solution is to create a compound key consisting of the fields Band Name, Date, and Start Time. An alternative solution is to create a compound primary key consisting of the fields Club Name, Date, and Start Time.

A way to avoid having a compound key would be to create a field called Booking Number. Each booking would get its own unique number, similar to an invoice number.

Here is another way to think about this event entity: Over time, a band plays in many clubs, and each club hires many bands. The BAND-to-CLUB relationship is, thus, a many-to-many relationship. Such relationships signal the need for a table between the two entities in the relationship. Here, you need the BOOKINGS table that associates the BAND and CLUB tables. An associative table is implemented by including the primary keys from the two tables that are associated. In this case, the primary keys from the BAND and CLUB tables are included as foreign keys in the BOOKINGS table.

Rule 5: Avoid data redundancy. You should not include extra (redundant) fields in a table. Doing this takes up extra disk space, and it leads to data entry errors because the same value must be entered in multiple tables, and the chance of a keystroke error increases. In large databases, keeping track of multiple instances of the same data is nearly impossible, and contradictory data entries become a problem.

Consider this example: Why wouldn't Club Phone Number be in the BOOKINGS table as a field? After all, the agent might have to call about some last-minute change for a booking and could quickly look up the number in the BOOKINGS table. Assume that the BOOKINGS table had Booking Number as the primary key and Club Phone Number as a field. Figure A-9 shows the BOOKINGS table with the unnecessary field.

BOOKINGS	
Field Name	***Data Type***
Booking Number (primary key)	Text
Band Name	Text
Club Name	Text
Club Phone Number	Text
Date	Date/Time
Start Time	Date/Time
End Time	Date/Time
Fee	Currency

Figure A-9 The BOOKINGS table with an unnecessary field—Club Phone Number

The fields Date, Start Time, End Time, and Fee logically depend on the Booking Number primary key—they help define the booking. Band Name and Club Name are foreign keys and are needed to establish the relationship between the tables BAND, CLUB, and BOOKINGS. But what about Club Phone Number? It is not defined by the Booking Number. It is defined by Club Name—*i.e., it's a function of the club, not of the booking.* Thus, the Club Phone Number field does not belong in the BOOKINGS table. It's already in the CLUB table, and if the agent needs it, he can look it up there.

Perhaps you can see the practical data entry problem with including Club Phone Number in BOOKINGS. Suppose that a club changed its contact phone number. The agent can easily change the number one time, in CLUB. But now the agent would need to remember the names of all the other tables that have that field as well, and change the values there too. Of course, with a small database, that might not be a difficult thing to recall. But in large databases having many redundant fields in many tables, this sort of maintenance becomes very difficult, which means that redundant data is often incorrect.

You might object, saying, "What about all those foreign keys? Aren't they redundant?" In a sense, they are. But they are needed to establish the relationship between one entity and another, as discussed previously.

Rule 6: Do not include a field if it can be calculated from other fields. A **calculated field** is made using the query generator. Thus, the agent's fee is not included in the BOOKINGS table because it can be calculated by query (here, 5% times the booking fee).

Practice Database Design Problem

Imagine this scenario: Your town has a library. The library wants to keep track of its business in a database, and you have been called in to build it. You talk to the town librarian, review the old paper-based records, and watch people use the library for a few days. You learn these things about the library:

1. Anyone who lives in the town can get a library card if they ask for one. The library considers each person who gets a card a "member" of the library.
2. The librarian wants to be able to contact members by telephone and by mail. She calls members if their books are overdue or when requested materials become available. She likes to mail a "thank you" note to each member on the yearly anniversary of their joining. Without a database, contacting members can be difficult to do efficiently; for example, there could be more than one member by the name of Sally Smith. Often, a parent and a child have the same first and last name, live at the same address, and share a phone.
3. The librarian tries to keep track of each member's reading "interests." When new books come in, the librarian alerts members whose interests match those books. For example, long-time member Sue Doaks is interested in Western novels, growing orchids, and baking bread. There must be some way to match such a reader's interests with available books. However, although the librarian wants to track all of a member's reading interests, she wants to classify each book as being in just one category of interest. For example, the classic gardening book *Orchids of France* would be classified as a book about orchids or a book about France, but not both.
4. The library stocks many books. Each book has a title and any number of authors. Conceivably, there could be more than one book in the library titled *History of the United States*. Similarly, there could be more than one author with the same name.
5. A writer could be the author of more than one book.
6. A book could be checked out repeatedly as time goes on. For example, *Orchids of France* could be checked out by one member in March, by another in July, and by yet another member in September.
7. The library must be able to identify whether a book is checked out.
8. A member can check out any number of books in a visit. Conceivably, a member could visit the library more than once a day to check out books, and some members do just that.
9. All books that are checked out are due back in two weeks, no exceptions. The "late" fee is 50 cents per day late. The librarian would like to have an automated way of generating an overdue book list each day, so she could telephone the miscreants.
10. The library has a number of employees. Each employee has a job title. The librarian is paid a salary, but other employees are paid by the hour. Employees clock in and clock out each day. Assume that all employees work only one shift per day, and all are paid weekly. Pay is deposited directly into employees' checking accounts—no checks are hand-delivered. The database needs to include the librarian and all other employees.

Design the library's database, following the rules set forth in this tutorial. Your instructor will specify the format for your work. Here are a few hints, in the form of questions:

- A book can have more than one author. An author can write more than one book. How would you describe the relationship between books and authors?
- The library lends books for free, of course. If you thought of checking out a book as a sale, for zero revenue, how would you handle the library's revenue-generating event?
- A member can check out any number of books in a check-out. A book can be checked out more than once. How would you describe the relationship between check-outs and books?

Microsoft Access Tutorial

Microsoft Access is a relational database package that runs on the Microsoft Windows operating system. This tutorial was prepared using Access 2003.

Before using this tutorial, you should know the fundamentals of Microsoft Access and know how to use Windows. This tutorial teaches you some advanced Access skills you'll need to do database case studies. This tutorial concludes with a discussion of common Access problems and how to solve them.

A preliminary caution: Always observe proper file-saving and closing procedures. Use these steps to exit from Access: (1) With your diskette in **Drive A:**, use these commands: File—Close, then (2) File—Exit. This gets you back to Windows. Always end your work with these two steps. Never pull out your diskette and walk away with work remaining on the screen, or you will lose your work.

To begin this tutorial, you will create a new database called **Employee**.

AT THE KEYBOARD

Open a new database (in the Task Pane—New—Blank database). (According to Microsoft, the Task Pane is a universal remote control, which saves the user steps.) Call the database **Employee**. If you are saving to a floppy diskette, first select the drive (**A:**), and then enter the filename. **EMPLOYEE.mdb** would be a good choice.

Your opening screen should resemble the screen shown in Figure B-1.

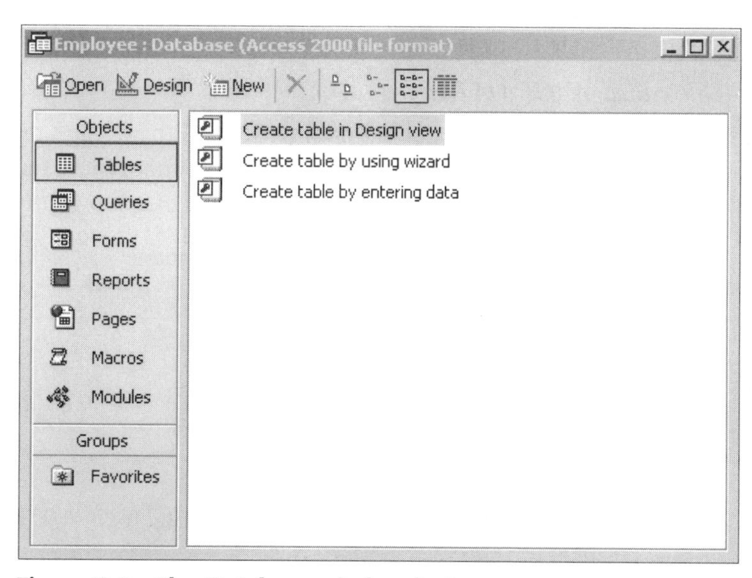

Figure B-1 The Database window in Access

In this tutorial, the screen shown in Figure B-1 is called the Database window. From this screen, you can create or change objects.

CREATING TABLES

Your database will contain data about employees, their wage rates, and their hours worked.

Defining Tables

In the Database window, make three new tables, using the instructions that follow.

AT THE KEYBOARD

(1) Define a table called EMPLOYEE.

This table contains permanent data about employees. To create it, in the Table Objects screen, click New, then Design View, and then define the table EMPLOYEE. The table's fields are Last Name, First Name, SSN (Social Security Number), Street Address, City, State, Zip, Date Hired, and US Citizens. The field SSN is the primary key field. Change the length of text fields from the default 50 spaces to more appropriate lengths; for example, the field Last Name might be 30 spaces, and the Zip field might be 10 spaces. Your completed definition should resemble the one shown in Figure B-2.

Table1 : Table

Field Name	Data Type	Description
Last Name	Text	
First Name	Text	
SSN	Text	
Street Address	Text	
City	Text	
State	Text	
Zip	Text	
Date Hired	Date/Time	
US Citizen	Yes/No	

Figure B-2 Fields in the EMPLOYEE table

When you're finished, choose File—Save. Enter the name desired for the table (here, EMPLOYEE). Make sure that you specify the name of the *table*, not the database itself. (Here, it is a coincidence that the EMPLOYEE table has the same name as its database file.)

(2) Define a table called WAGE DATA.

This table contains permanent data about employees and their wage rates. The table's fields are SSN, Wage Rate, and Salaried. The field SSN is the primary key field. Use the data types shown in Figure B-3. Your definition should resemble the one shown in Figure B-3.

Table1 : Table

Field Name	Data Type	Description
SSN	Text	
Wage Rate	Currency	
Salaried	Yes/No	

Figure B-3 Fields in the WAGE DATA table

Use File—Save to save the table definition. Name the table WAGE DATA.

(3) Define a table called HOURS WORKED.

The purpose of this table is to record the number of hours employees work each week in the year. The table's fields are SSN (text), Week # (number—long integer), and Hours (number—double). The SSN and Week# are the compound keys.

In the following example, the employee having SSN 089-65-9000 worked 40 hours in Week 1 of the year and 52 hours in Week 2.

SSN	Week #	Hours
089-65-9000	1	40
089-65-9000	2	52

Note that no single field can be the primary key field. Why? Notice that 089-65-9000 is an entry for each week. If the employee works each week of the year, at the end of the year, there will be 52 records with that value. Thus, SSN values will not distinguish records. However, no other single field can distinguish these records either, because other employees will have worked during the same week number, and some employees will have worked the same number of hours (40 would be common).

However, a table must have a primary key field. The solution? Use a compound primary key; that is, use values from more than one field. Here, the compound key to use consists of the field SSN plus the Week # field. Why? There is only *one* combination of SSN 089-65-9000 and Week# 1—those values *can occur in only one record*; therefore, the combination distinguishes that record from all others.

How do you set a compound key? The first step is to highlight the fields in the key. These must appear one after the other in the table definition screen. (Plan ahead for this format.) Alternately, you can highlight one field, hold down the Control key, and highlight the next.

AT THE KEYBOARD

For the HOURS WORKED table, click in the first field's left prefix area, hold the button down, then drag down to highlight names of all fields in the compound primary key. Your screen should resemble the one shown in Figure B-4.

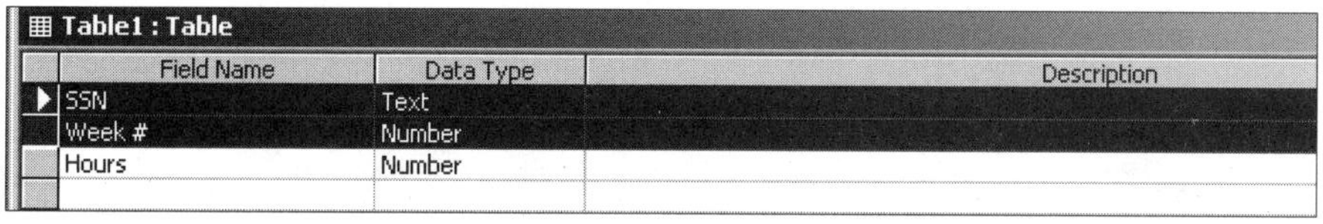

Table1 : Table

Field Name	Data Type	Description
SSN	Text	
Week #	Number	
Hours	Number	

Figure B-4 Selecting fields as the compound primary key for the HOURS WORKED table

Now, click the Key icon. Your screen should resemble the one shown in Figure B-5.

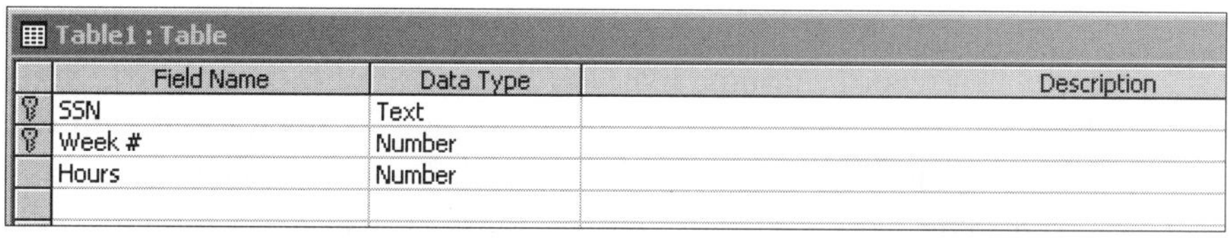

Table1 : Table

Field Name	Data Type	Description
SSN	Text	
Week #	Number	
Hours	Number	

Figure B-5 The compound primary key for the HOURS WORKED table

That completes the compound primary key and the table definition. Use File—Save to save the table as HOURS WORKED.

Adding Records to a Table

At this point, all you have done is to set up the skeletons of three tables. The tables have no data records yet. If you were to print out the tables, all you would see would be column headings (the field names). The most direct way to enter data into a table is to select the table, open it, and type the data directly into the cells.

AT THE KEYBOARD

At the Database window, select Tables, then EMPLOYEE. Then select Open. Your data entry screen should resemble the one shown in Figure B-6.

Employee : Table

Last Name	First Name	SSN	Street Address	City	State	Zip	Date Hired	US Citizen

Figure B-6 The data entry screen for the EMPLOYEE table

The table has many fields, and some of them may be off the screen, to the right. Scroll to see obscured fields. (Scrolling happens automatically as data is entered.) Figure B-6 has been adjusted to view all fields on one screen.

Type in your data, one field value at a time. Note that the first row is empty when you begin. Each time you finish a value, hit Enter, and the cursor will move to the next cell. After the last cell in a row, the cursor moves to the first cell of the next row, *and* Access automatically saves the record. (Thus, there is no File—Save step after entering data into a table.)

Dates (e.g., Date Hired) are entered as "6/15/04" (without the quotation marks). Access automatically expands the entry to the proper format in output.

Yes/No variables are clicked (checked) for Yes; otherwise (for No), the box is left blank. You can click the box from Yes to No, as if you were using a toggle switch.

If you make errors in data entry, click in the cell, backspace over the error, and type the correction.

Enter the data shown in Figure B-7 into the EMPLOYEE table.

Employee : Table

Last Name	First Name	SSN	Street Address	City	State	Zip	Date Hired	US Citizen
Howard	Jane	114-11-2333	28 Sally Dr	Glasgow	DE	19702	8/1/2005	☑
Smith	John	123-45-6789	30 Elm St	Newark	DE	19711	6/1/1996	☑
Smith	Albert	148-90-1234	44 Duce St	Odessa	DE	19722	7/15/1987	☑
Jones	Sue	222-82-1122	18 Spruce St	Newark	DE	19716	7/15/2004	☐
Ruth	Billy	714-60-1927	1 Tater Dr	Baltimore	MD	20111	8/15/1999	☐
Add	Your	Data	Here	Newark	MN	33776		☑

Figure B-7 Data for the EMPLOYEE table

Note that the sixth record is *your* data record. The edit pencil in the left prefix area marks that record. Assume that you live in Newark, Minnesota, were hired on today's date (enter the date), and are a U.S. citizen. (Later in this tutorial, you will see that one entry is for the author's name and the SSN 099-11-3344 for this record.)

Open the WAGE DATA table and enter the data shown in Figure B-8 into the table.

Wage Data : Table

SSN	Wage Rate	Salaried
114-11-2333	$10.00	☐
123-45-6789	$0.00	☑
148-90-1234	$12.00	☐
222-82-1122	$0.00	☑
714-60-1927	$0.00	☑
Your SSN!	$8.00	☐

Figure B-8 Data for the WAGE DATA table

Tutorial B

Again, you must enter your SSN. Assume that you earn $8 an hour and are not salaried. (Note that Salaried = No implies someone is paid by the hour. Those who are salaried do not get paid by the hour, so their hourly rate is shown as 0.00.)

Open the HOURS WORKED table and enter the data shown in Figure B-9 into the table.

Hours Worked : Table

SSN	Week #	Hours
114-11-2333	1	40
114-11-2333	2	50
123-45-6789	1	40
123-45-6789	2	40
148-90-1234	1	38
148-90-1234	2	40
222-82-1122	1	40
222-82-1122	2	40
714-60-1927	1	40
714-60-1927	2	40
your SSN	1	60
your SSN	2	55

Figure B-9 Data for the HOURS WORKED table

Notice that salaried employees are always given 40 hours. Non-salaried employees (including you) might work any number of hours. For your record, enter your SSN, 60 hours worked for Week 1, and 55 hours worked for Week 2.

CREATING QUERIES

Because you can already create basic queries, this section teaches you the kinds of advanced queries you will create in the Case Studies.

Using Calculated Fields in Queries

A **calculated field** is an output field that is made from *other* field values. A calculated field is *not* a field in a table; it is created in the query generator. The calculated field does not become part of the table—it is just part of query output. The best way to explain this process is by working through an example.

AT THE KEYBOARD

Suppose that you want to see the SSNs and wage rates of hourly workers, and you want to see what the wage rates would be if all employees were given a 10% raise. To do this, show the SSN, the current wage rate, and the higher rate (which should be titled New Rate in the output). Figure B-10 shows how to set up the query.

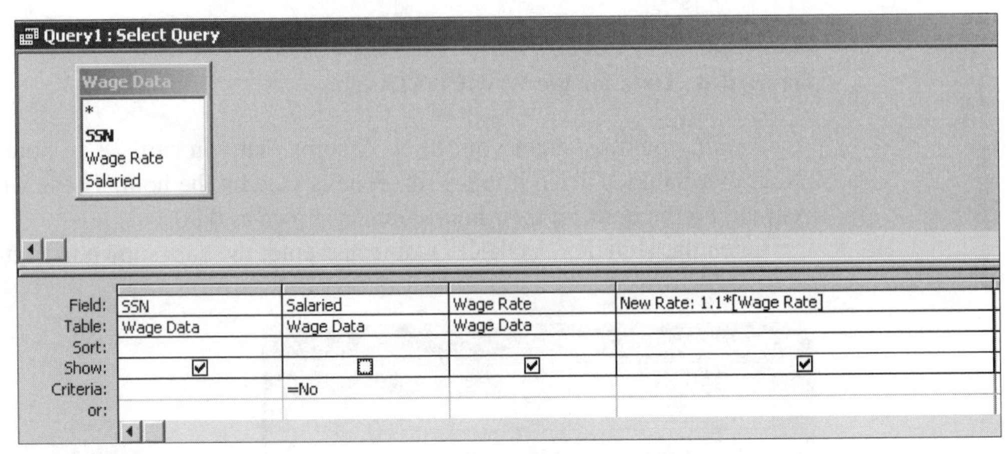

Figure B-10 Query set-up for the calculated field

The Salaried field is needed, with the Criteria =No, to select hourly workers. The Show box for that field is not checked, so the Salaried field values will not show in the query output.

Note the expression for the calculated field, which you see in the right-most field cell:

New Rate: 1.1*[Wage Rate]

New Rate: merely specifies the desired output heading. (Don't forget the colon.) The 1.1*[Wage Rate] multiplies the old wage rate by 110%, which results in the 10% raise.

In the expression, the field name Wage Rate must be enclosed in square brackets. This is a rule: *Any time that an Access expression refers to a field name, it must be enclosed in square brackets.*

If you run this query, your output should resemble that shown in Figure B-11.

Query1 : Select Query

SSN	Wage Rate	New Rate
114-11-2333	$10.00	11
148-90-1234	$12.00	13.2
099-11-3344	$8.00	8.8

Figure B-11 Output for a query with calculated field

Notice that the calculated field output is not shown in Currency format; it's shown as a Double—a number with digits after the decimal point. To convert the output to Currency format, click the line above the calculated field expression, thus activating the column (it darkens). Your data entry screen should resemble the one shown in Figure B-12.

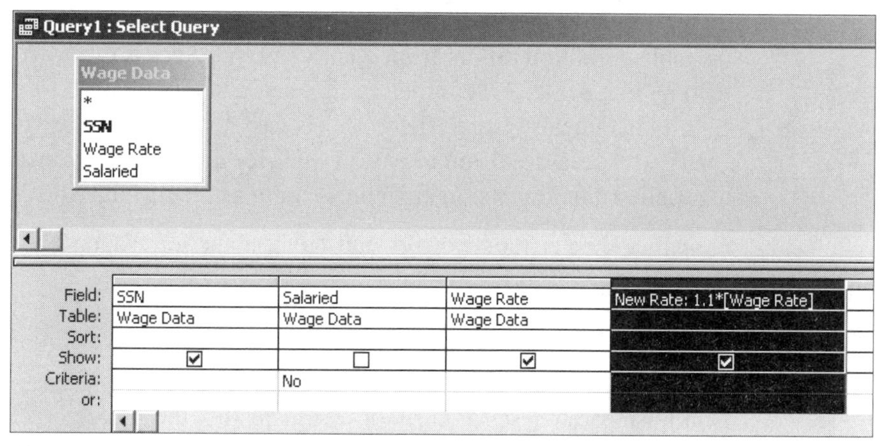

Figure B-12 Activating a calculated field in query design

Then select View—Properties. Click the Format drop-down menu. A window, such as the one shown in Figure B-13, will pop up.

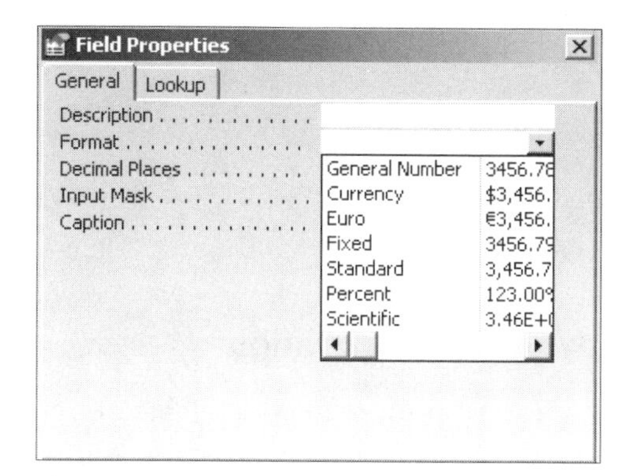

Figure B-13 Field Properties of a calculated field

Click Currency. Then click the upper-right X to close the window. Now when you run the query, the output should resemble that shown in Figure B-14.

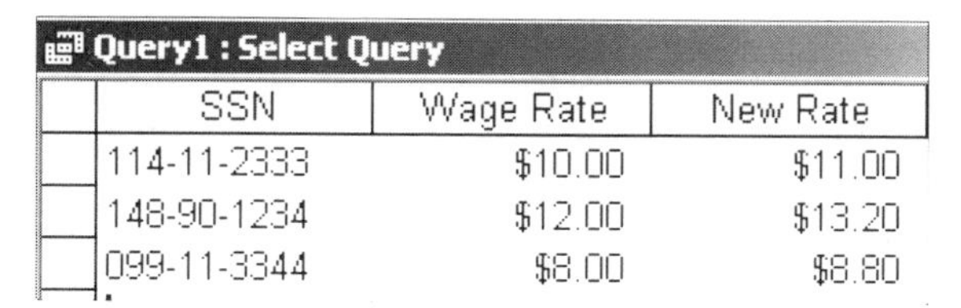

Query1 : Select Query

SSN	Wage Rate	New Rate
114-11-2333	$10.00	$11.00
148-90-1234	$12.00	$13.20
099-11-3344	$8.00	$8.80

Figure B-14 Query output with formatted calculated field

Next, let's look at how to avoid errors when making calculated fields.

Avoiding Errors in Making Calculated Fields

Follow these guidelines to avoid making errors in calculated fields:

- Don't put the expression in the *Criteria* cell, as if the field definition were a filter. You are making a field, so put the expression in the *Field* cell.

- Spell, capitalize, and space a field's name *exactly* as you did in the table definition. If the table definition differs from what you type, Access thinks you're defining a new field by that name. Access then prompts you to enter values for the new field, which it calls a "Parameter Query" field. This is easy to debug because of the tag Parameter Query. If Access asks you to enter values for a Parameter, you almost certainly have misspelled a field name in an expression in a calculated field or a criterion.

  Example: Here are some errors you might make for Wage Rate:

  Misspelling: (Wag Rate)

  Case change: (wage Rate / WAGE RATE)

  Spacing change: (WageRate / Wage   Rate)

- Don't use parentheses or curly braces instead of the square brackets. Also, don't put parentheses inside square brackets. You *are* allowed to use parentheses outside the square brackets, in the normal algebraic manner.

  Example: Suppose that you want to multiply Hours times Wage Rate, to get a field called Wages Owed. This is the correct expression:

  Wages Owed: [Wage Rate]*[Hours]

  This would also be correct:

  Wages Owed: ([Wage Rate]*[Hours])

  But it would **not** be correct to leave out the inside brackets, which is a common error:

  Wages Owed: [Wage Rate*Hours]

"Relating" Two (or More) Tables by the Join Operation

Often, the data you need for a query is in more than one table. To complete the query, you must join the tables. One rule of thumb is that joins are made on fields that have common *values,* and those fields can often be key fields. The names of the join fields are irrelevant—the names may be the same, but that is not a requirement for an effective join.

Make a join by first bringing in (Adding) the tables needed. Next, decide which fields you will join. Then, click one field name and hold down the left mouse button while dragging the cursor over to the other field's name in its window. Release the button. Access puts a line in, signifying the join. (*Note*: If there are two fields in the tables with the same name, Access will put in the line automatically, so you do not have to do the click-and-drag operation.)

You can join more than two tables together. The common fields *need not* be the same in all tables; that is, you can "daisy chain" them together.

A common join error is to Add a table to the query and then fail to link it to another table. You have a table just "floating" in the top part of the QBE screen! When you run the query, your output will show the same records over and over. This error is unmistakable because there is *so much* redundant output. The rules are (1) add only the tables you need, and (2) link all tables.

Next, you'll work through an example of a query needing a join.

At the Keyboard

Suppose that you want to see the last names, SSNs, wage rates, salary status, and citizenship only for U.S. citizens and hourly workers. The data is spread across two tables, EMPLOYEE and WAGE DATA, so both tables are added, and five fields are pulled down. Criteria are then added. Set up your work to resemble that shown in Figure B-15.

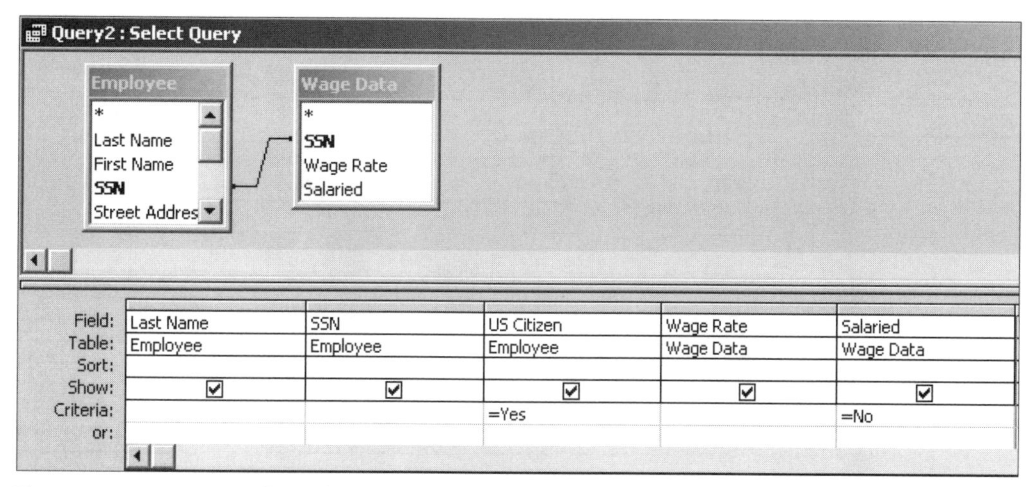

Figure B-15 A query based on two joined tables

In Figure B-15, the join is on the SSN field. A field by that name is in both tables, so Access automatically puts in the join. If one field had been spelled SSN and the other Social Security Number, you would still join on these fields (because of the common values). You would click and drag to do this operation.

Now run the query. The output should resemble that shown in Figure B-16, with the exception of the name Brady.

Query2 : Select Query

Last Name	SSN	US Citizen	Wage Rate	Salaried
Howard	114-11-2333	☑	$10.00	☐
Smith	148-90-1234	☑	$12.00	☐
Brady	099-11-3344	☑	$8.00	☐

Figure B-16 Output of a query based on two joined tables

Here is a quick review of Criteria: If you want data for employees who are U.S. citizens *and* who are hourly workers, the Criteria expressions go into the *same* Criteria row. If you want data for employees who are U.S. citizens *or* who are hourly workers, one of the expressions goes into the second Criteria row (the one that has the "or:" notation).

There is no need to print the query output or to save it. Go back to the Design View and close the query. Another practice query follows.

AT THE KEYBOARD

Suppose that you want to see the wages owed to hourly employees for Week 2. Show the last name, the SSN, the salaried status, week #, and the wages owed. Wages will have to be a calculated field ([Wage Rate]*[Hours]). The criteria are =No for Salaried and =2 for the Week #. (Another "And" query!) You'd set up the query the way it is displayed in Figure B-17.

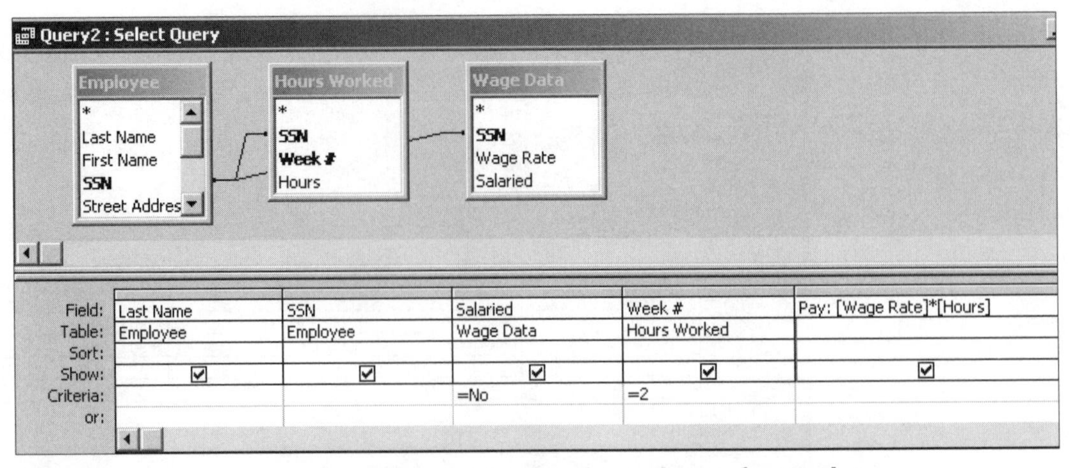

Figure B-17 Query set-up for wages owed to hourly employees for Week 2

In the previous table, the calculated field column was widened so you can see the whole expression. To widen a column, remember to click on the column boundary line and drag to the right.

Run the query. The output should be similar to that shown in Figure B-18 (if you formatted your calculated field to currency).

Query2 : Select Query

Last Name	SSN	Salaried	Week #	Pay
Howard	114-11-2333	☐	2	$500.00
Smith	148-90-1234	☐	2	$480.00
Brady	099-11-3344	☐	2	$440.00

Figure B-18 Query output for wages owed to hourly employees for Week 2

Notice that it was not necessary to pull down the Wage Rate and Hours fields to make this query work. Return to the Design View. There is no need to save. Select File—Close.

Summarizing Data from Multiple Records (Sigma Queries)

You may want data that summarizes values from a field for several records (or possibly all records) in a table. For example, you might want to know the average hours worked for all employees in a week, or perhaps the total (sum of) all the hours worked. Furthermore, you might want data grouped ("stratified") in some way. For example, you might want to know the average hours worked, grouped by all U.S. citizens versus all non-U.S. citizens. Access calls this kind of query a "summary" query, or a **Sigma query**. Unfortunately, this terminology is not intuitive, but the statistical operations that are allowed will be familiar. These operations include the following:

Sum	The total of some field's values
Count	A count of the number of instances in a field, i.e., the number of records. Here, to get the number of employees, you'd count the number of SSN numbers.
Average	The average of some field's values

Min	The minimum of some field's values
Var	The variance of some field's values
StDev	The standard deviation of some field's values

AT THE KEYBOARD

Suppose that you want to know how many employees are represented in a database. The first step is to bring the EMPLOYEE table into the QBE screen. Do that now. The query will Count the number of SSNs, which is a Sigma query operation. Thus, you must bring down the SSN field.

To tell Access you want a Sigma query, click the little "Sigma" icon in the menu, as shown in Figure B-19.

Figure B-19 Sigma icon

This opens up a new row in the lower part of the QBE screen, called the Total row. At this point, the screen would resemble that shown in Figure B-20.

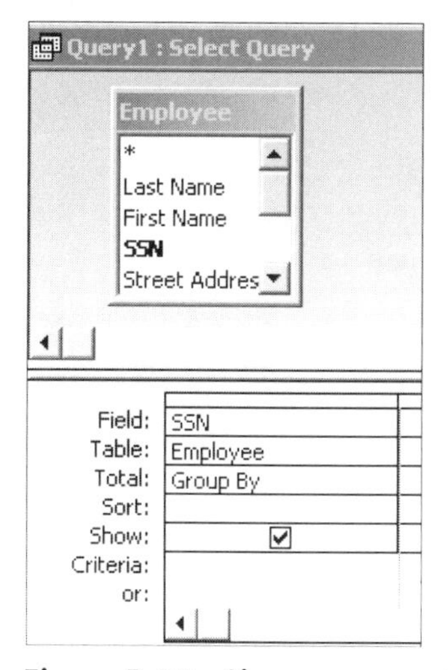

Figure B-20 Sigma query set-up

Note that the Total cell contains the words "Group By." Until you specify a statistical operation, Access just assumes that a field will be used for grouping (stratifying) data.

To count the number of SSNs, click next to Group By, revealing a little arrow. Click the arrow to reveal a drop-down menu, as shown in Figure B-21.

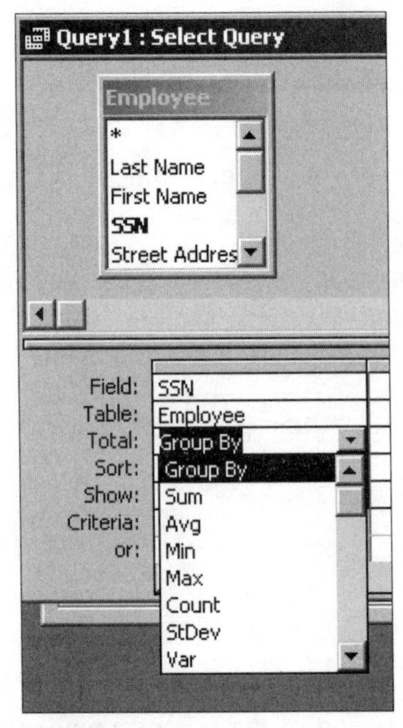

Figure B-21 Choices for statistical operation in a Sigma query

Select the Count operator. (With this menu, you may need to scroll to see the operator you want.) Your screen should now resemble that shown in Figure B-22.

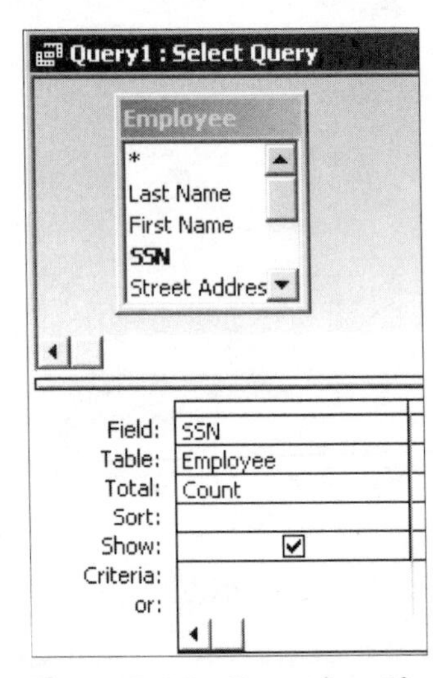

Figure B-22 Count in a Sigma query

Run the query. Your output should resemble that shown in Figure B-23.

Figure B-23 Output of Count in a Sigma query

Notice that Access has made a pseudo-heading "CountOfSSN." To do this, Access just spliced together the statistical operation (Count), the word *Of*, and the name of the field

(SSN). What if you wanted an English phrase, such as, "Count of Employees," as a heading? In the Design View, you'd change the query to resemble the one shown in Figure B-24.

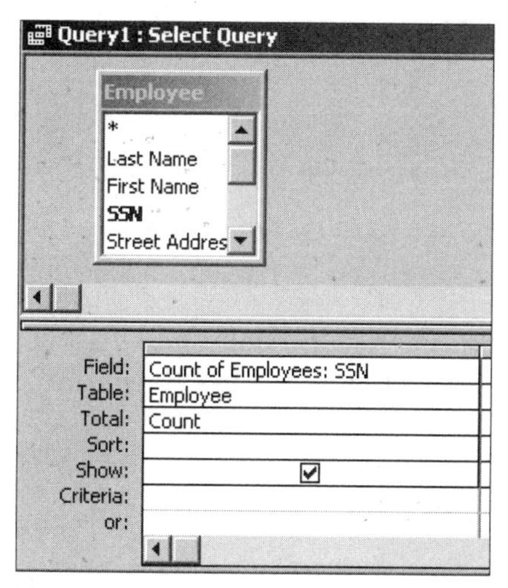

Figure B-24 Heading change in a Sigma query

Now when you run the query, the output should resemble that shown in Figure B-25.

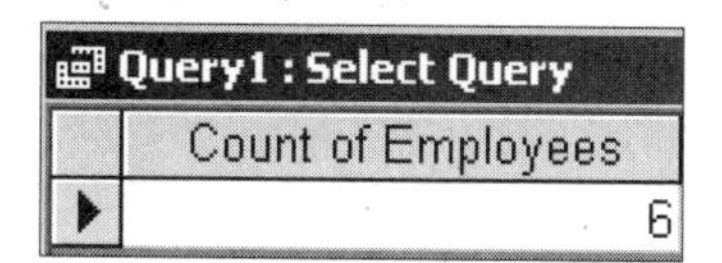

Figure B-25 Output of heading change in a Sigma query

There is no need to save this query. Go back to the Design View and Close.

At the Keyboard

Here is another example. Suppose that you want to know the average wage rate of employees, grouped by whether they are salaried.

Figure B-26 shows how your query should be set up.

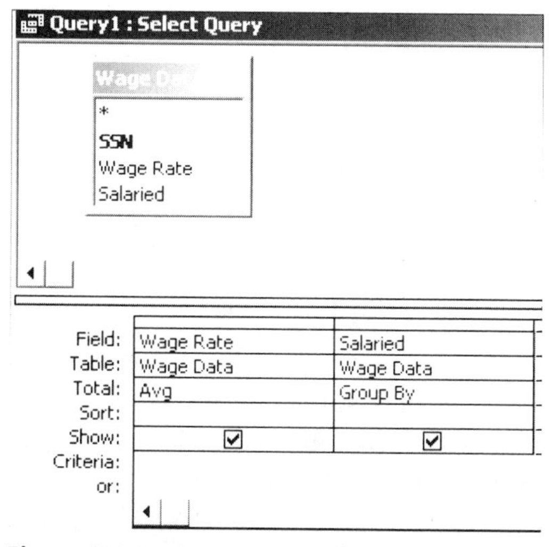

Figure B-26 Query set-up for average wage rate of employees

When you run the query, your output should resemble that shown in Figure B-27.

Query1 : Select Query

AvgOfWage Rate	Salaried
$0.00	☑
$10.00	☐

Figure B-27 Output of query for average wage rate of employees

Recall the convention that salaried workers are assigned zero dollars an hour. Suppose that you want to eliminate the output line for zero dollars an hour because only hourly-rate workers matter for this query. The query set-up is shown in Figure B-28.

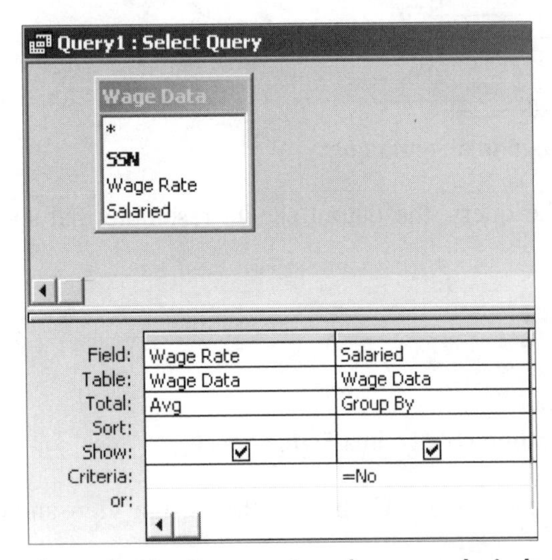

Figure B-28 Query set-up for non-salaried workers only

When you run the query, you'll get output for non-salaried employees only, as shown in Figure B-29.

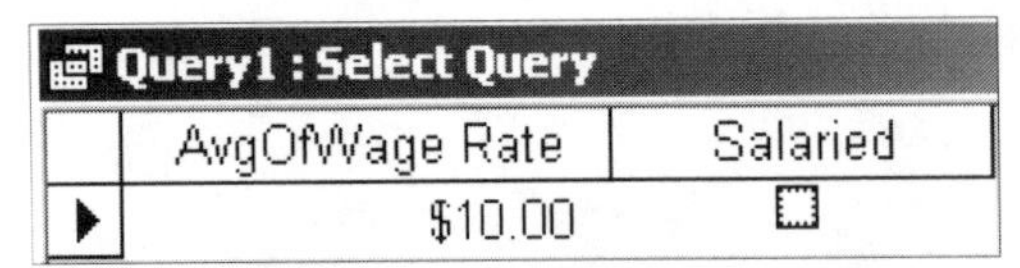
Query1 : Select Query

AvgOfWage Rate	Salaried
$10.00	☐

Figure B-29 Query output for non-salaried workers only

Thus, it's possible to use a Criteria in a Sigma query without any problem, just as you would with a "regular" query.

There is no need to save the query. Go back to the Design View and Close.

At the Keyboard

You can make a calculated field in a Sigma query. Assume that you want to see two things for hourly workers: (1) the average wage rate—call it Average Rate in the output; and (2) 110% of this average rate—call it the Increased Rate.

You already know how to do certain things for this query. The revised heading for the average rate will be Average Rate (Average Rate: Wage Rate, in the Field cell). You want the Average of that field. Grouping would be by the Salaried field (with Criteria: =No, for hourly workers).

The most difficult part of this query is to construct the expression for the calculated field. Conceptually, it is as follows:

Increased Rate: 1.1*[The current average, however that is denoted]

The question is how to represent [The current average]. You cannot use Wage Rate for this, because that heading denotes the wages before they are averaged. Surprisingly, it turns out that you can use the new heading (Average Rate) to denote the averaged amount. Thus:

Increased Rate: 1.1*[Average Rate]

Thus, counterintuitively, *you can treat "Average Rate" as if it were an actual field name.* Note, however, that if you use a calculated field, such as Average Rate, in another calculated field, as shown in Figure B-30, you must show that original calculated field in the query output, or the query will ask you to "enter parameter value," which is incorrect. Use the set-up shown in Figure B-30.

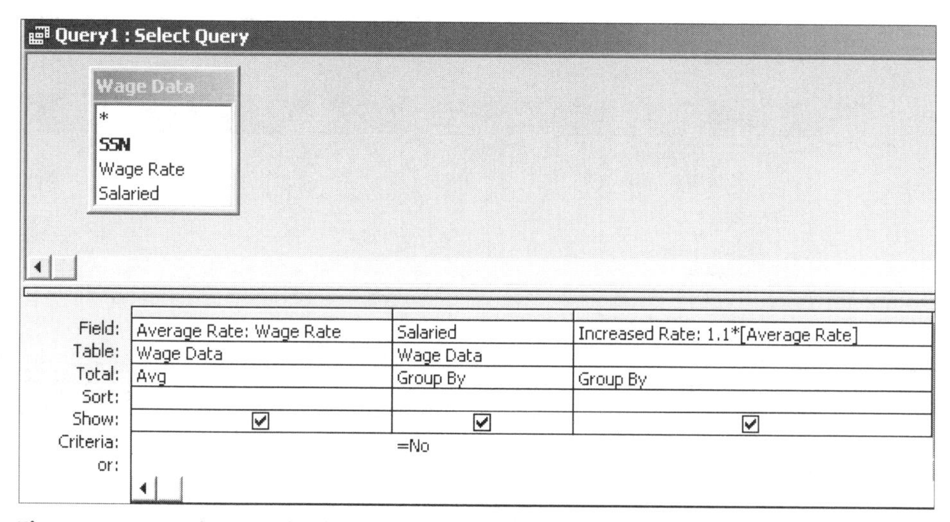

Figure B-30 Using a calculated field in another calculated field

However, if you ran the query now shown in Figure B-30, you'd get some sort of error message. You do not want Group By in the calculated field's Total cell. There is not a *statistical* operator that applies to the calculated field. You must change the Group By operator to Expression. You may have to scroll to get to Expression in the list. Figure B-31 shows how your screen should look.

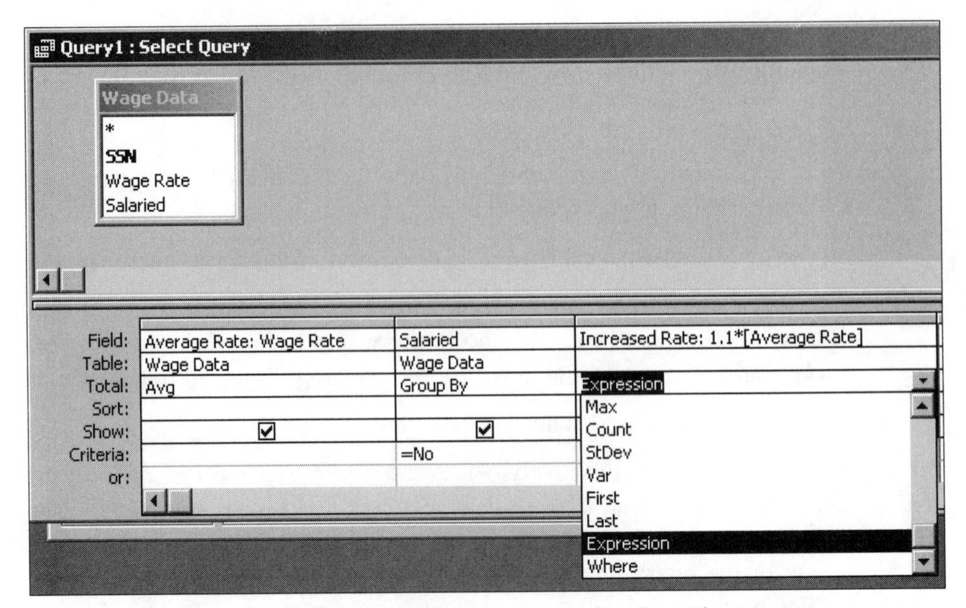

Figure B-31 Changing the Group By to an Expression in a Sigma query

Figure B-32 shows how the screen looks before running the query.

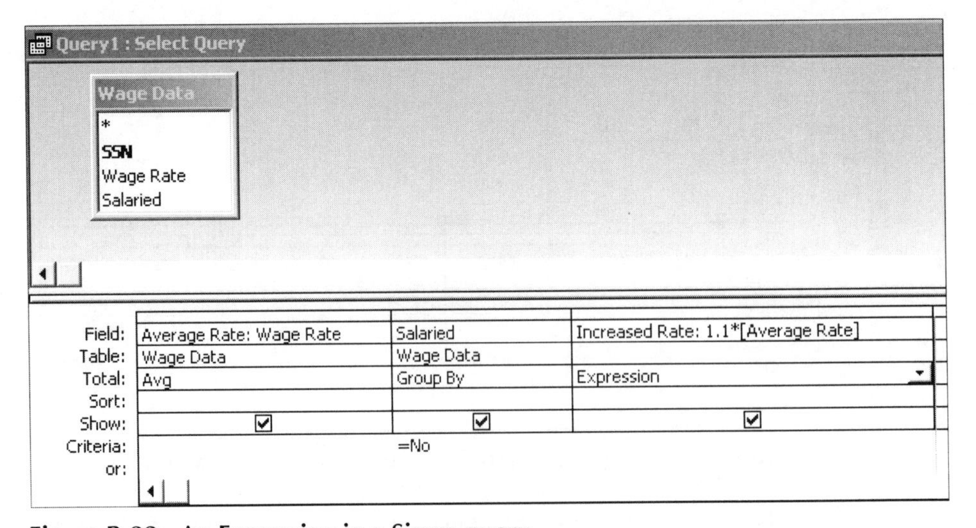

Figure B-32 An Expression in a Sigma query

Figure B-33 shows the output of the query.

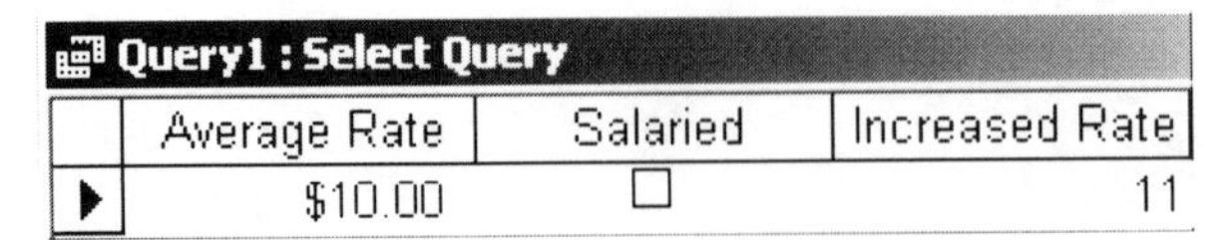

Figure B-33 Output of an Expression in a Sigma query

There is no need to save the query definition. Go back to the Design View. Select File—Close.

Using the Date() Function in Queries

Access has two date function features that you should know about. A description of them follows.

1. The following built-in function gives you *today's date*:

    Date()

    You can use this function in a query criteria or in a calculated field. The function "returns" the day on which the query is run. (I.e., it puts that value into the place where the function is in an expression.)

2. *Date arithmetic* lets you subtract one date from another to obtain the number of days difference. Access would evaluate the following expression as the integer 5 (9 less 4 is 5).

    10/9/2004 – 10/4/2004

Here is an example of how date arithmetic works. Suppose that you want to give each employee a bonus equaling a dollar for each day the employee has worked for you. You'd need to calculate the number of days between the employee's date of hire and the day that the query is run, then multiply that number by 1.

The number of elapsed days is shown by the following equation:

Date() – [Date Hired]

Suppose that for each employee, you want to see the last name, SSN, and bonus amount. You'd set up the query as shown in Figure B-34.

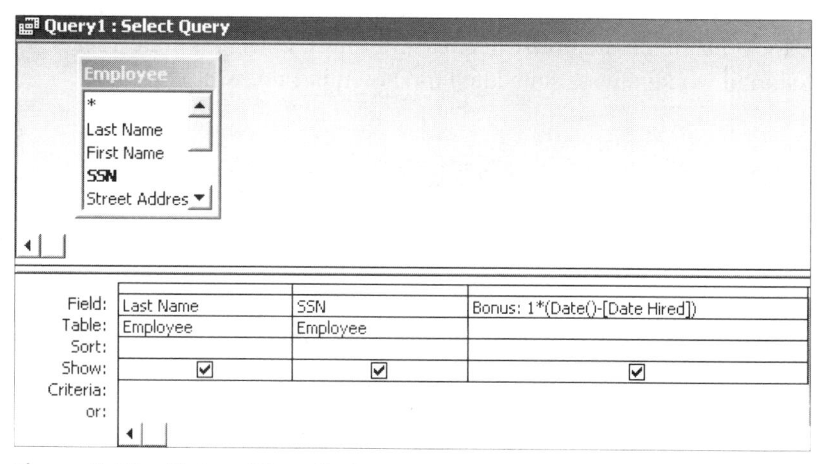

Figure B-34 Date arithmetic in a query

Assume that you set the format of the Bonus field to Currency. The output will be similar to Figure B-35. (Your Bonus data will be different because you are working on a date different than the date when this tutorial was written.)

Query1 : Select Query

Last Name	SSN	Bonus
Brady	099-11-3344	$0.00
Howard	114-11-2333	$144.00
Smith	123-45-6789	$2,396.00
Smith	148-90-1234	$5,640.00
Jones	222-82-1122	$526.00
Ruth	714-60-1927	$1,226.00

Figure B-35 Output of query with date arithmetic

Using Time Arithmetic in Queries

Access will also let you subtract the values of time fields to get an elapsed time. Assume that your database has a JOB ASSIGNMENTS table showing the times that non-salaried employees were at work during a day. The definition is shown in Figure B-36.

Job Assignments : Table

Field Name	Data Type
SSN	Text
ClockIn	Date/Time
ClockOut	Date/Time
Date	Date/Time

Figure B-36 Date/Time data definition in the JOB ASSIGNMENTS table

Assume that the Date field is formatted for Long Date and that the ClockIn and ClockOut fields are formatted for Medium Time. Assume that, for a particular day, non-salaried workers were scheduled as shown in Figure B-37.

Job Assignments : Table

SSN	ClockIn	ClockOut	Date
099-11-3344	8:30 AM	4:30 PM	Friday, September 30, 2005
114-11-2333	9:00 AM	3:00 PM	Friday, September 30, 2005
148-90-1234	7:00 AM	5:00 PM	Friday, September 30, 2005

Figure B-37 Display of date and time in a table

You want a query that will show the elapsed time on premises for the day. When you add the tables, your screen may show the links differently. Click and drag the JOB ASSIGNMENTS, EMPLOYEE, and WAGE DATA table icons to look like those in Figure B-38.

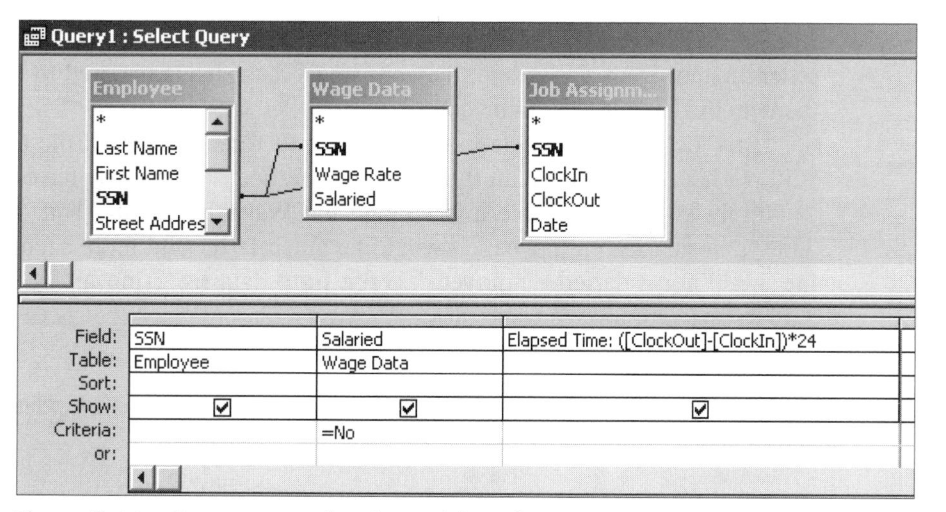

Figure B-38 Query set-up for time arithmetic

Figure B-39 shows the output.

Query1 : Select Query

SSN	Salaried	Elapsed Time
114-11-2333	☐	6
148-90-1234	☐	10
099-11-3344	☐	8

Figure B-39 Query output for time arithmetic

The output looks right. For example, employee 099-11-3344 was at work from 8:30 a.m. to 4:30 p.m., which is eight hours. But how does the odd expression that follows yield the correct answers?

([ClockOut] – [ClockIn]) * 24

Why wouldn't the following expression, alone, work?

[ClockOut] – [ClockIn]

This is the answer: In Access, *subtracting one time from the other yields the decimal portion of a 24-hour day*. Employee 099-11-3344 worked 8 hours, which is a third of a day, so .3333 would result. That is why you must multiply by 24—to convert to an hour basis. Continuing with 099-11-3344, 1/3 x 24 = 8.

Note that parentheses are needed to force Access to do the subtraction *first*, before the multiplication. Without parentheses, multiplication takes precedence over subtraction. With the following expression, ClockIn would be multiplied by 24 and then that value would be subtracted from ClockOut, and the output would be a nonsense decimal number:

[ClockOut] – [ClockIn] * 24

Delete and Update Queries

Thus far, the queries presented in this tutorial have been Select queries. They select certain data from specific tables, based on a given criterion. You can also create queries to update the original data in a database. Businesses do this often, and in real time. For example, when you order an item from a Web site, the company's database is updated to reflect the purchase of the item by deleting it from inventory.

Let's look at an example. Suppose that you want to give all the non-salaried workers a $.50 per hour pay raise. With the three non-salaried workers you have now, it would be easy to simply go into the table and change the Wage Rate data. But assume that you have 3,000 non-salaried employees. It would be much faster and more accurate to change each of the 3,000 non-salaried employees' Wage Rate data by using an Update query to add the $.50 to each employee's wage rate.

AT THE KEYBOARD

Let's change each of the non-salaried employees' pay via an Update query. Figure B-40 shows how to set up the query.

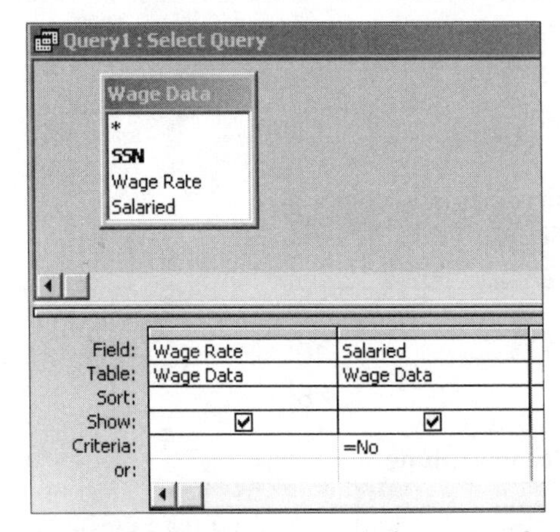

Figure B-40 Query set-up for an Update Query

So far, this query is just a Select query. Place your cursor somewhere above the QBE grid, and then right-click the mouse. Once in that menu, choose Query Type—Update Query, as shown in Figure B-41.

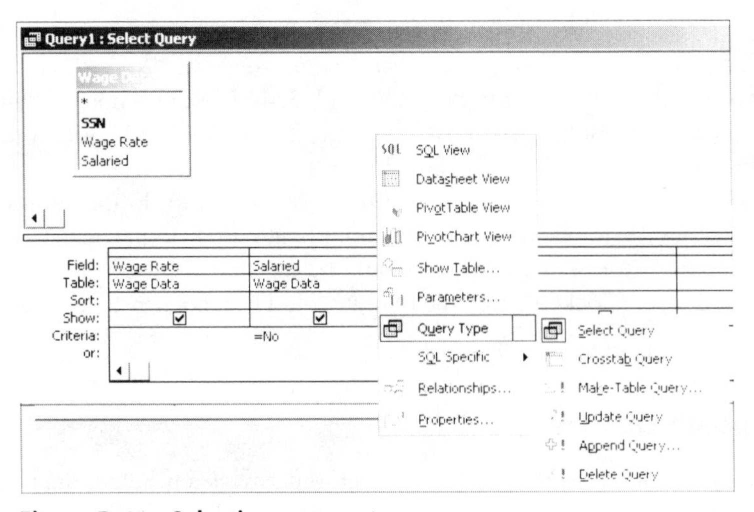

Figure B-41 Selecting a query type

Notice that you now have another line on the QBE grid called "Update to:" This is where you specify the change or update to the data. Notice that you are going to update only the non-salaried workers by using a filter under the Salaried field. Update the Wage Rate data to Wage Rate plus $.50, as shown in Figure B-42 . (Note the [ ] as in a calculated field.)

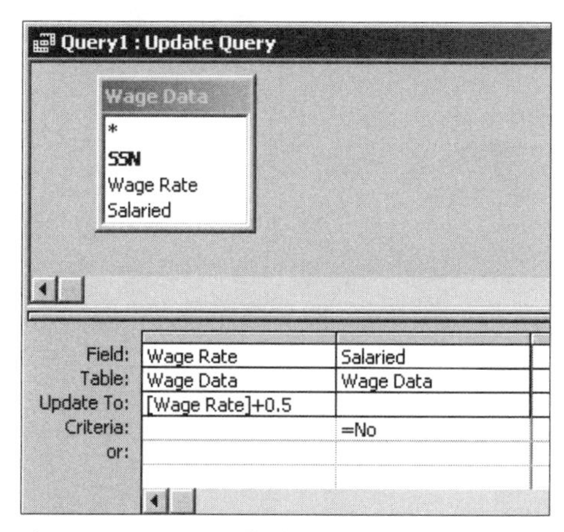

Figure B-42 Updating the wage rate for non-salaried workers

Now run the query. You will first get a warning message, as shown in Figure B-43.

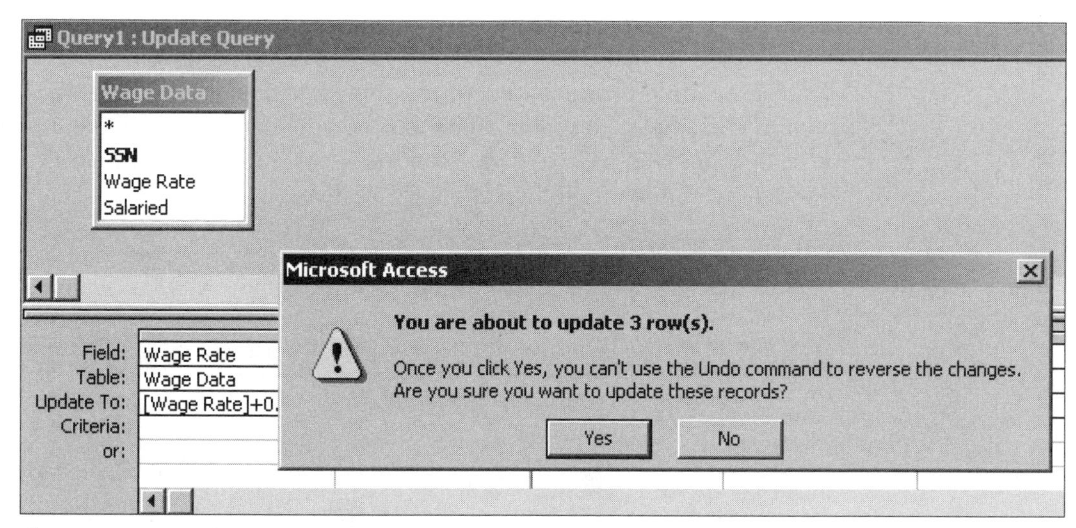

Figure B-43 Update Query warning

Once you click "Yes," the records will be updated. Check those updated records now by viewing the WAGE DATA table. Each salaried wage rate should now be increased by $.50. Note that in this example, you are simply adding $.50 to each salaried wage rate. You could add or subtract data from another table as well. If you do that, remember to call the field name in square brackets.

Delete queries work the same way as Update queries. Assume that your company has been taken over by the state of Delaware. The state has an odd policy of only employing Delaware residents. Thus, you must delete (or fire) all employees who are not Delaware residents. To do this, you would first create a Select query using the EMPLOYEE table, right-click your mouse, choose Delete Query from Query Type, then bring down the State field and filter only those records not in Delaware (DE). Do not perform this operation, but note that, if you did, the set-up would look like that in Figure B-44.

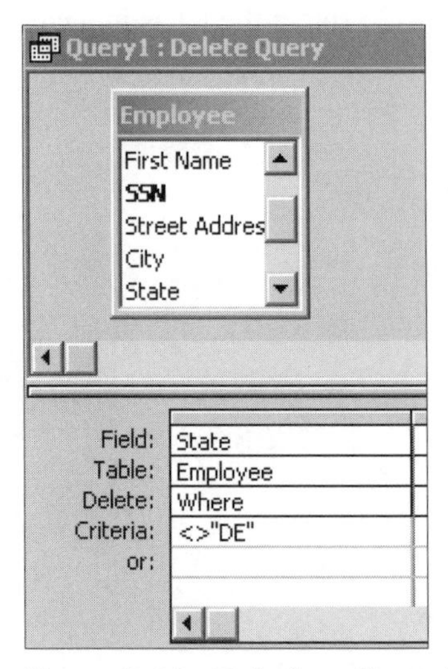

Figure B-44 Deleting all employees who are not Delaware residents

Parameter Queries

Another type of query, which is a type of Select query, is called a **Parameter query**. Here is an example: Suppose that your company has 5,000 employees. You might want to query the database to find the same kind of information again and again, only about different employees. For example, you might want to query the database to find out how many hours a particular employee has worked. To do this, you could run a query previously created and stored, but run it only for a particular employee.

At the keyboard

Create a Select query with the format shown in Figure B-45.

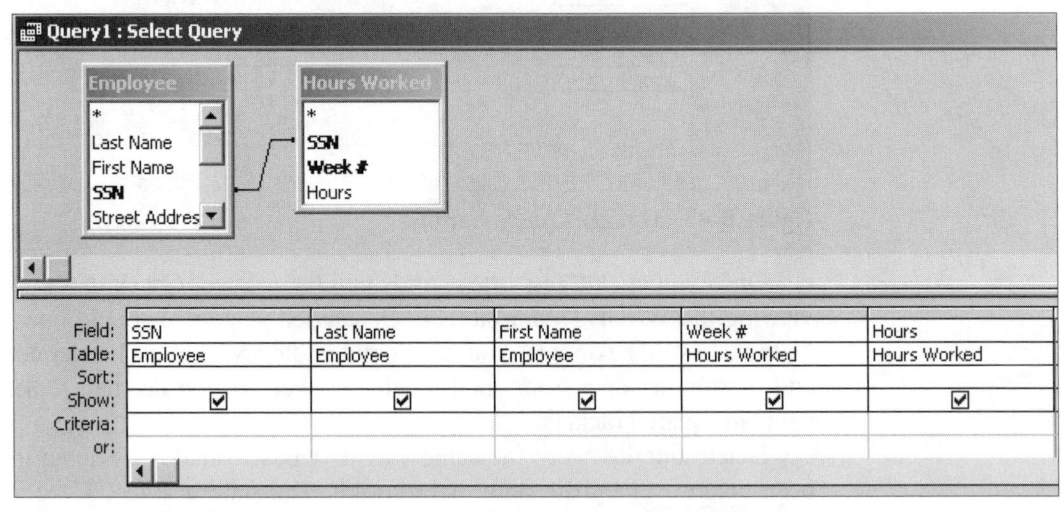

Figure B-45 Design of a Parameter query begins as a Select query

In the Criteria line of the QBE grid for the field SSN, type in what is shown in Figure B-46.

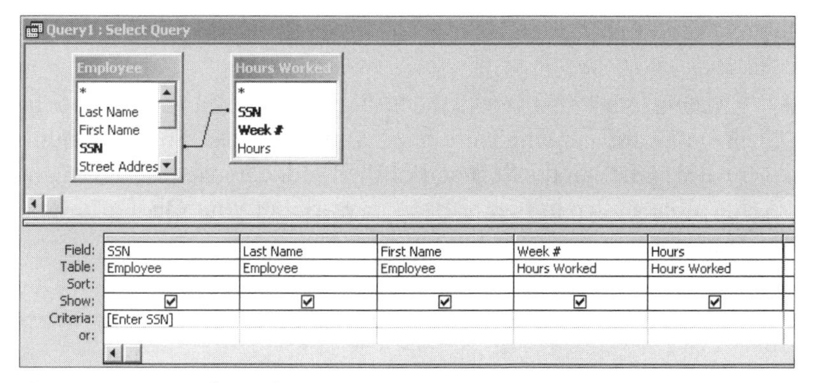

Figure B-46 Design of a Parameter query

Note the square brackets, as you would expect to see in a calculated field.

Now run that query. You will be prompted for the specific employee's SSN, as shown in Figure B-47.

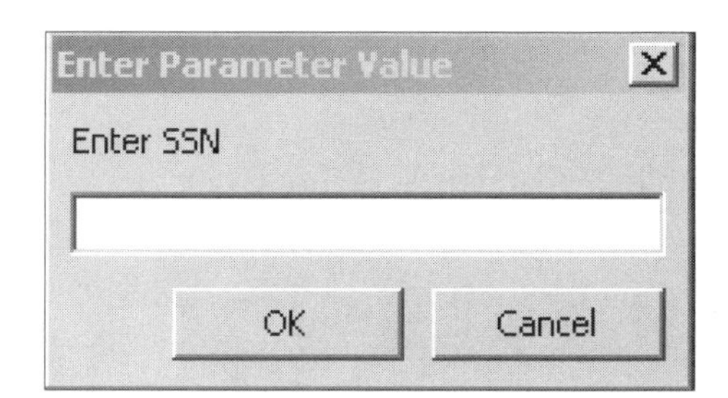

Figure B-47 Enter Parameter Value dialog box

Type in your own SSN. Your query output should resemble that shown in Figure B-48.

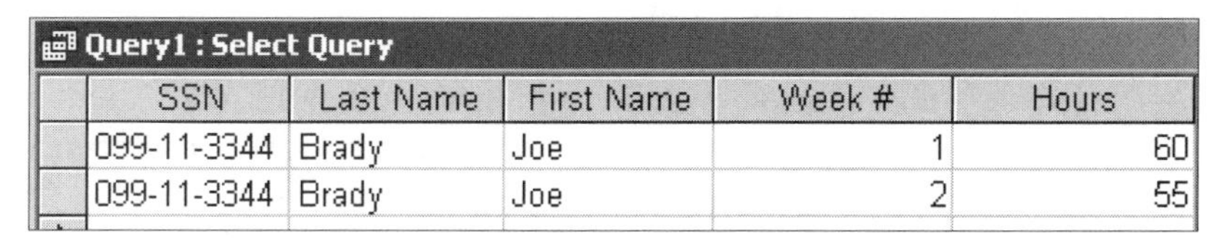

Query1 : Select Query

SSN	Last Name	First Name	Week #	Hours
099-11-3344	Brady	Joe	1	60
099-11-3344	Brady	Joe	2	55

Figure B-48 Output of a Parameter query

Seven Practice Queries

This portion of the tutorial is designed to provide you with additional practice in making queries. Before making these queries, you must create the specified tables and enter the records shown in the Creating Tables section of this tutorial. The output shown for the practice queries is based on those inputs.

At the Keyboard

For each query that follows, you are given a problem statement and a "scratch area." You are also shown what the query output should look like. Follow this procedure: Set up a query in Access. Run the query. When you are satisfied with the results, save the query and continue with the next query. You will be working with the EMPLOYEE, HOURS WORKED, and WAGE DATA tables.

1. Create a query that shows the SSN, last name, state, and date hired for those living in Delaware *and* who were hired after 12/31/92. Sort (ascending) by SSN. (Sorting review: Click in the Sort cell of the field. Choose Ascending or Descending.) Use the table shown in Figure B-49 to work out your QBE grid on paper before creating your query.

Field					
Table					
Sort					
Show					
Criteria					
Or:					

Figure B-49 QBE grid template

Your output should resemble that shown in Figure B-50.

Query 1 : Select Query

SSN	Last Name	State	Date Hired
114-11-2333	Howard	DE	8/1/2005
123-45-6789	Smith	DE	6/1/1996
222-82-1122	Jones	DE	7/15/2004

Figure B-50 Number 1 query output

2. Create a query that shows the last name, first name, date hired, and state for those living in Delaware *or* who were hired after 12/31/92. The primary sort (ascending) is on last name, and secondary sort (ascending) is on first name. (Review: The Primary Sort field must be left of the Secondary Sort field in the query set-up.) Use the table shown in Figure B-51 to work out your QBE grid on paper before creating your query.

Field					
Table					
Sort					
Show					
Criteria					
Or:					

Figure B-51 QBE grid template

If your name were Brady, your output would look like that shown in Figure B-52.

Query 1 : Select Query

Last Name	First Name	Date Hired	Stat
Brady	Joe	12/23/2002	MN
Howard	Jane	8/1/2005	DE
Jones	Sue	7/15/2004	DE
Ruth	Billy	8/15/1999	MD
Smith	Albert	7/15/1987	DE
Smith	John	6/1/1996	DE

Figure B-52 Number 2 query output

3. Create a query that shows the sum of hours worked by U.S. citizens and by non-U.S. citizens (i.e., group on citizenship). The heading for total hours worked should be Total Hours Worked. Use the table shown in Figure B-53 to work out your QBE grid on paper before creating your query.

Field					
Table					
Total					
Sort					
Show					
Criteria					
Or:					

Figure B-53 QBE grid template

Your output should resemble that shown in Figure B-54.

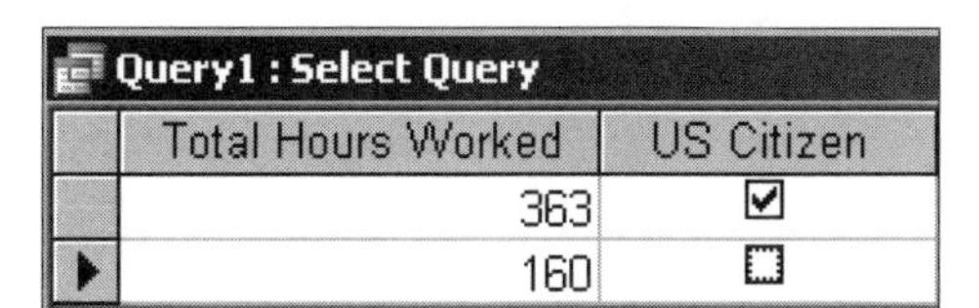

Query1 : Select Query

Total Hours Worked	US Citizen
363	☑
160	☐

Figure B-54 Number 3 query output

4. Create a query that shows the wages owed to hourly workers for Week 1. The heading for the wages owed should be Total Owed. The output headings should be: Last Name, SSN, Week #, and Total Owed. Use the table shown in Figure B-55 to work out your QBE grid on paper before creating your query.

Field					
Table					
Sort					
Show					
Criteria					
Or:					

Figure B-55 QBE grid template

If your name were Joseph Brady, your output would look like that in Figure B-56.

Query1 : Select Query

Last Name	SSN	Week #	Total Owed
Howard	114-11-2333	1	$420.00
Smith	148-90-1234	1	$475.00
Brady	099-11-3344	1	$510.00

Figure B-56 Number 4 query output

5. Create a query that shows the last name, SSN, hours worked, and overtime amount owed for employees paid hourly who earned overtime during Week 2. Overtime is paid at 1.5 times the normal hourly rate for hours over 40. The amount shown should be just the overtime portion of the wages paid. This is not a Sigma query—amounts should be shown for individual workers. Use the table shown in Figure B-57 to work out your QBE grid on paper before creating your query.

Field					
Table					
Sort					
Show					
Criteria					
Or:					

Figure B-57 QBE grid template

If your name were Joseph Brady, your output would look like that shown in Figure B-58.

Query1 : Select Query

Last Name	SSN	Hours	OT Pay
Howard	114-11-2333	50	$157.50
Brady	099-11-3344	55	$191.25
*			

Figure B-58 Number 5 query output

6. Create a Parameter query that shows the hours employees have worked. Have the Parameter query prompt for the week number. The output headings should be Last Name, First Name, Week #, and Hours. Do this only for the non-salaried workers. Use the table shown in Figure B-59 to work out your QBE grid on paper before creating your query.

Field					
Table					
Sort					
Show					
Criteria					
Or:					

Figure B-59 QBE grid template

Run the query with "2" when prompted for the Week #. Your output should look like that shown in Figure B-60.

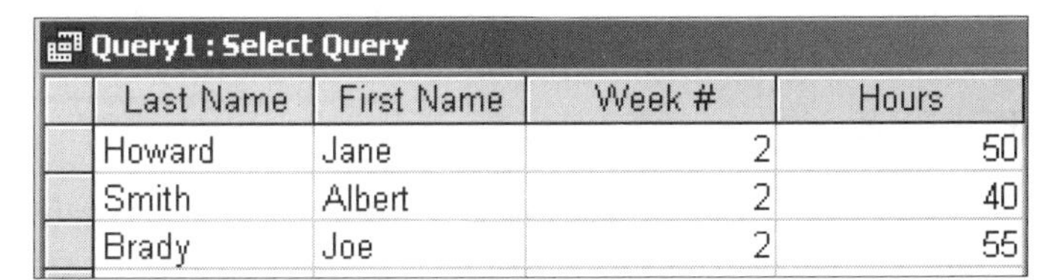

Query1 : Select Query

Last Name	First Name	Week #	Hours
Howard	Jane	2	50
Smith	Albert	2	40
Brady	Joe	2	55

Figure B-60 Number 6 query output

7. Create an update query that gives certain workers a merit raise. You must first create an additional table as shown in Figure B-61.

Merit Raises : Table

SSN	Merit Raise
114-11-2333	$0.25
148-90-1234	$0.15

Figure B-61 MERIT RAISES table

Now make a query that adds the Merit Raise to the current Wage Rate for those who will receive a raise. When you run the query, you should be prompted with "You are about to update two rows." Check the original WAGE DATA table to confirm the update. Use the table shown in Figure B-62 to work out your QBE grid on paper before creating your query.

Field					
Table					
Update to					
Criteria					
Or:					

Figure B-62 QBE grid template

CREATING REPORTS

Database packages let you make attractive management reports from a table's records or from a query's output. If you are making a report from a table, the Access report generator looks up the data in the table and puts it into report format. If you are making a report from a query's output, Access runs the query in the background (you do not control this or see this happen) and then puts the output in report format.

There are three ways to make a report. One is to handcraft the report in the Design View, from scratch. This is tedious and is not shown in this tutorial. The second way is to use the Report Wizard, during which Access leads you through a menu-driven construction. This method is shown in this tutorial. The third way is to start in the Wizard and then use the Design View to tailor what the Wizard produces. This method is also shown in this tutorial.

Creating a Grouped Report

This tutorial assumes that you can use the Wizard to make a basic ungrouped report. This section of the tutorial teaches you how to make a grouped report. (If you cannot make an ungrouped report, you might learn how to make one by following the first example that follows.)

AT THE KEYBOARD

Suppose that you want to make a report out of the HOURS WORKED table. At the main Objects menu, start a new report by choosing Reports—New. Select the Report Wizard and select the HOURS WORKED table from the drop-down menu as the report basis. Select OK. In the next screen, select all the fields (using the >> button), as shown in Figure B-63.

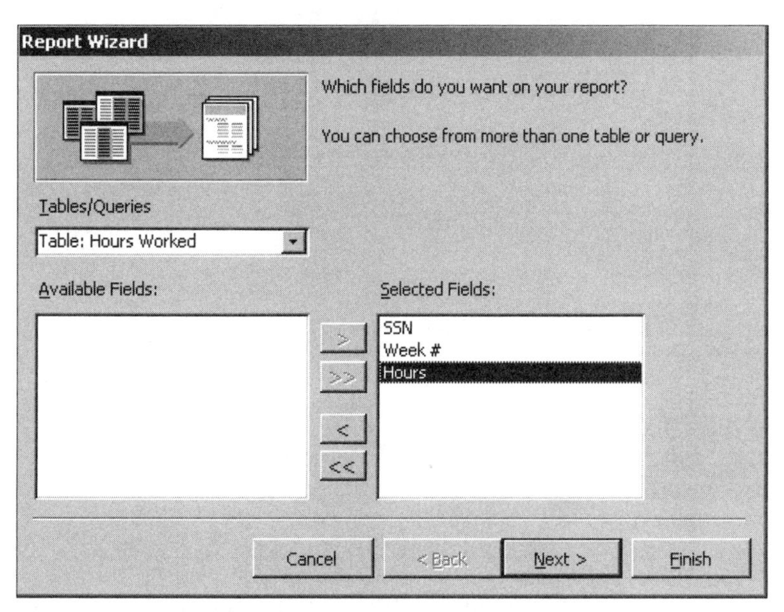

Figure B-63 Field selection step in the Report Wizard

Click Next. Then tell Access that you want to group on Week # by double-clicking that field name. You'll see that shown in Figure B-64.

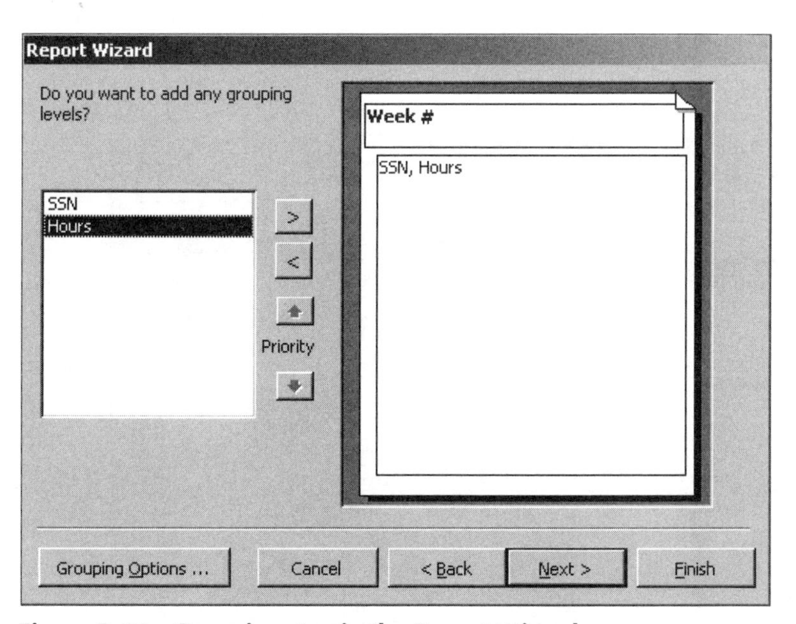

Figure B-64 Grouping step in the Report Wizard

Click Next. You'll see a screen, similar to the one shown in Figure B-65, for Sorting and for Summary Options.

Tutorial B

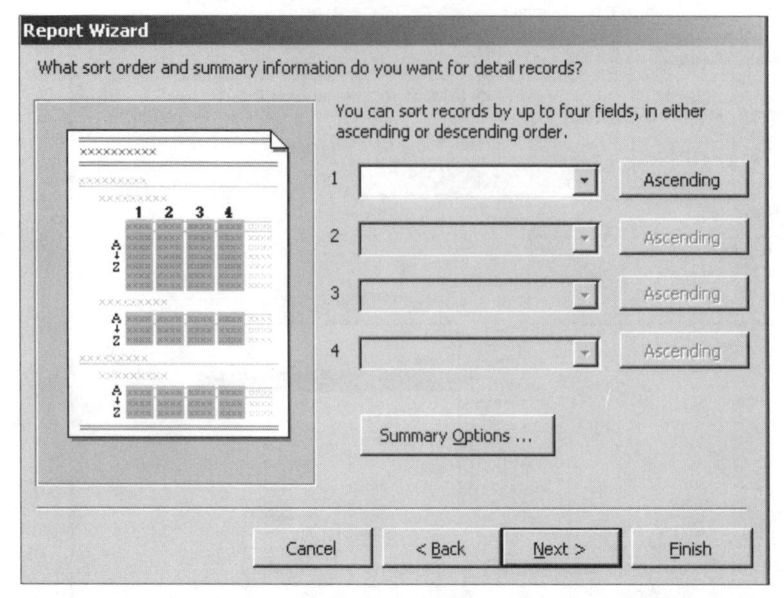

Figure B-65 Sorting and Summary Options step in the Report Wizard

Because you chose a grouping field, Access will now let you decide whether you want to see group subtotals and/or report grand totals. All numeric fields could be added, if you choose that option. In this example, group subtotals are for total hours in each week. Assume that you *do* want the total of hours by week. Click Summary Options. You'll get a screen similar to the one in Figure B-66.

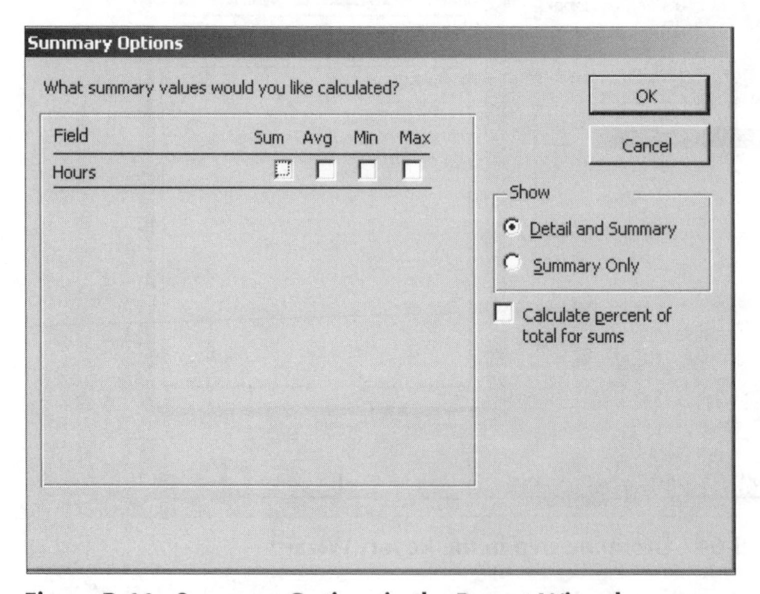

Figure B-66 Summary Options in the Report Wizard

Next, follow these steps:

1. Click the Sum box for Hours (to sum the hours in the group).
2. Click Detail and Summary. (Detail equates with "group," and Summary with "grand total for the report.")
3. Click OK. This takes you back to the Sorting screen, where you can choose an ordering within the group, if desired. (In this case, none is.)

4. Click Next to continue.
5. In the Layout screen (not shown here) choose Stepped and Portrait.
6. Make sure that the "Adjust the field width so all fields fit on a page" check box is unchecked.
7. Click Next.
8. In the Style screen (not shown), accept Corporate.
9. Click Next.
10. Provide a title—Hours Worked by Week would be appropriate.
11. Select the Preview button to view the report.
12. Click Finish.

The top portion of your report will look like that shown in Figure B-67.

Hours Worked by Week

Week #	*SSN*	*Hours*
1		
	099-11-3344	60
	714-60-1927	40
	222-82-1122	40
	148-90-1234	38
	123-45-6789	40
	114-11-2333	40
Summary for 'Week #' = 1 (6 detail records)		
Sum		258

Figure B-67 Hours Worked by Week report

Notice that data is shown grouped by weeks, with Week 1 on top, then a subtotal for that week. Week 2 data is next, then there is a grand total (which you can scroll down to see). The subtotal is labeled "Sum," which is not very descriptive. This can be changed later in the Design View. Also, there is the apparently useless italicized line that starts out *"Summary for 'Week ..."* This also can be deleted later in the Design View. At this point, you should select File—Save As (accept the suggested title if you like). Then select File—Close to get back the Database window. Try it. Your report's Objects screen should resemble that shown in Figure B-68.

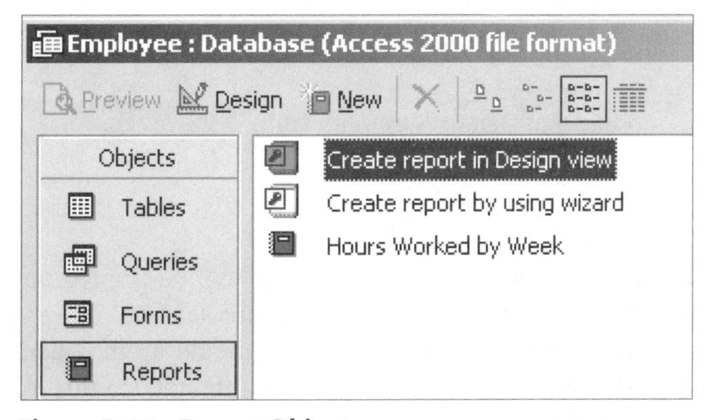

Figure B-68 Report Objects screen

To edit the report in the Design View, click the report title, then the Design button. You will see a complex (and intimidating) screen, similar to the one shown in Figure B-69.

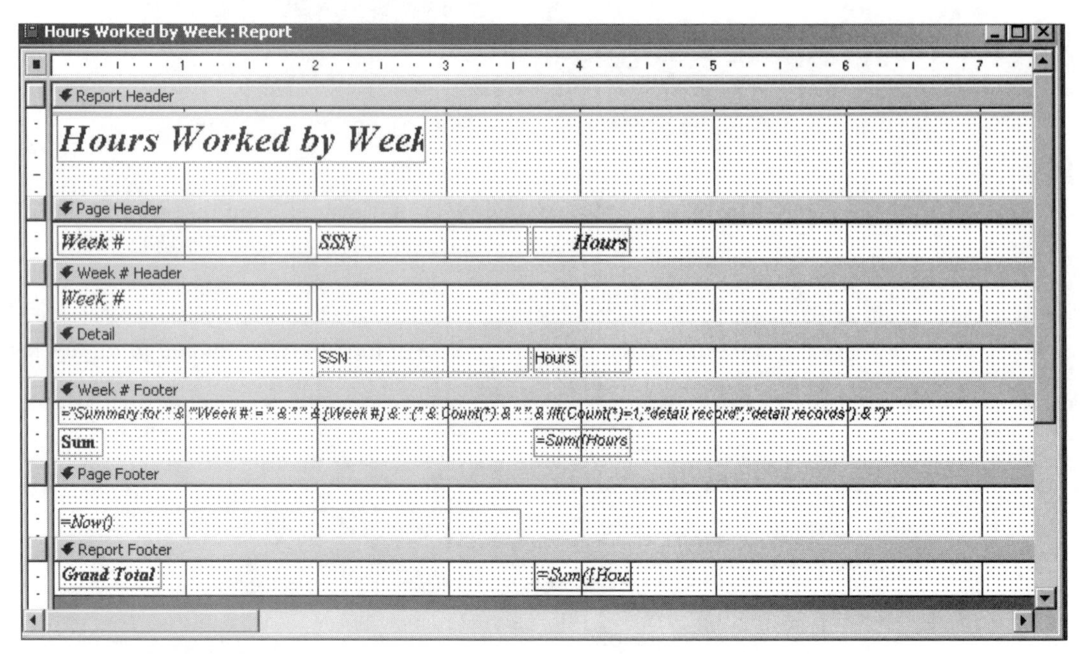

Figure B-69 Report design screen

The organization of the screen is hierarchical. At the top is the Report level. The next level down (within a report) is the Page level. The next level or levels down (within a page) are for any data groupings you have specified.

If you told Access to make group (summary) totals, your report will have a Report Header area and end with a Grand Total in the Report Footer. The report header is usually just the title you have specified.

A page also has a header, which is usually just the names of the fields you have told Access to put in the report (here, Week #, SSN, and Hours fields). Sometimes the page number is put in by default.

Groupings of data are more complex. There is a header for the group—in this case, the *value* of the Week # will be the header; for example, there is a group of data for the first week, then one for the second—the values shown will be 1 and 2. Within each data grouping is the other "detail" that you've requested. In this case, there will be data for each SSN and the related hours.

Each Week # gets a "footer," which is a labeled sum—recall that you asked for that to be shown (Detail and Summary were requested). The Week # Footer is indicated by three things:

1. The italicized line that starts =Summary for ...
2. The Sum label
3. The adjacent expression =Sum(Hours)

The italicized line beneath the Week # Footer will be printed unless you eliminate it. Similarly, the word "Sum" will be printed as the subtotal label unless you eliminate it. The "=Sum(Hours)" is an expression that tells Access to add up the quantity *for the header in question* and put that number into the report as the subtotal. (In this example, that would be the sum of hours, by Week #.)

Each report also gets a footer—the grand total (in this case, of hours) for the report.

If you look closely, each of the detail items appears to be doubly inserted in the design. For example, you will see the notation for SSN twice, once in the Page Header and then again in the Detail band. Hours are treated similarly.

The data items will not actually be printed twice, because each data element is an object in the report; each object is denoted by a label and by its value. There is a representation of the name, which is the boldface name itself (in this example, "SSN" in the page header), and there is a representation in less-bold type for the value "SSN" in the Detail band.

Sometimes, the Report Wizard is arbitrary about where it puts labels and data. However, if you do not like where the Wizard puts data, the objects containing data can be moved around in the Design View. You can click and drag within the band or across bands. Often, a box will be too small to allow full numerical values to show. When that happens, select the box and then click one of the sides to stretch it. This will allow full values to show. At other times, an object's box will be very long. When that happens, the box can be clicked, re-sized, then dragged right or left in its panel to reposition the output.

Suppose that you do *not* want the italicized line to appear in the report. Also suppose that you would like different subtotal and grand total labels. The italicized line is an object that can be activated by clicking it. Do that. "Handles" (little squares) appear around its edges, as shown in Figure B-70.

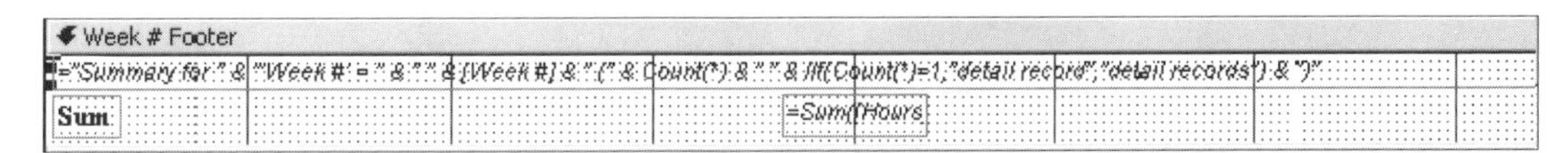

Figure B-70 Selecting an object in the Report Design View

Press the Delete key to get rid of the selected object.

To change the subtotal heading, click the Sum object, as shown in Figure B-71.

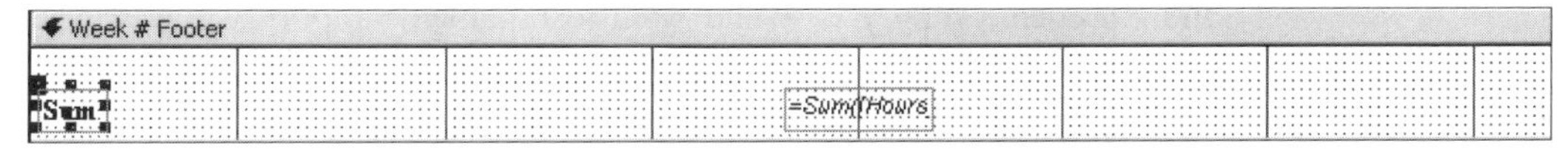

Figure B-71 Selecting the Sum object in the Report Design View

Click again. This gives you an insertion point from which you can type, as shown in Figure B-72.

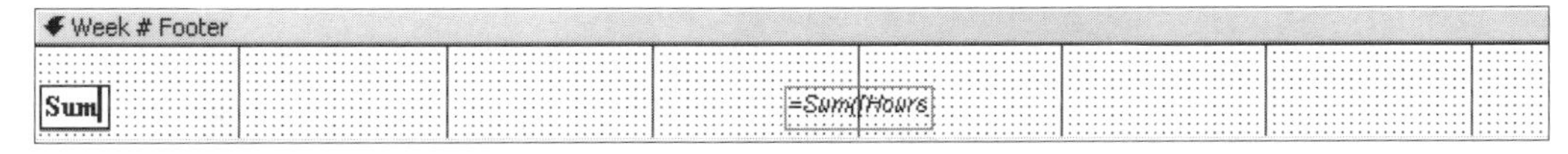

Figure B-72 Typing in an object in the Report Design View

Change the label to something like Sum of Hours for Week, then hit Enter, or click somewhere else in the report to deactivate. Your screen should resemble that shown in Figure B-73.

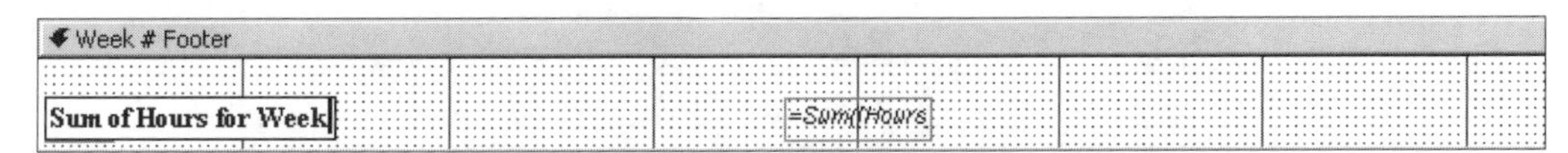

Figure B-73 Changing a label in the Report Design View

You can change the Grand Total in the same way.

Finally, you'll want to save and then print the file: File—Save. Then select File—Print Preview. You should see a report similar to that in Figure B-74 (top part is shown).

Hours Worked by Week

Week #	*SSN*	*Hours*
1		
	099-11-3344	60
	714-60-1927	40
	222-82-1122	40
	148-90-1234	38
	123-45-6789	40
	114-11-2333	40
Sum of Hours for Week		258

Figure B-74 Hours Worked by Week report

Notice that the data are grouped by week number (data for Week 1 is shown) and subtotaled for that week. The report would also have a grand total at the bottom.

Moving Fields in the Design View

When you group on more than one field in the Report Wizard, the report has an odd "staircase" look. There is a way to overcome that effect in the Design View, which you will learn next.

Suppose that you make a query showing an employee's last name, street address, zip code and wage rate. Then you make a report from that query, grouping on last name, street address, and zip code. (Why you would want to organize a report in this way is not clear, but for the moment, accept the organization for the purpose of the example.) This is shown in Figure B-75.

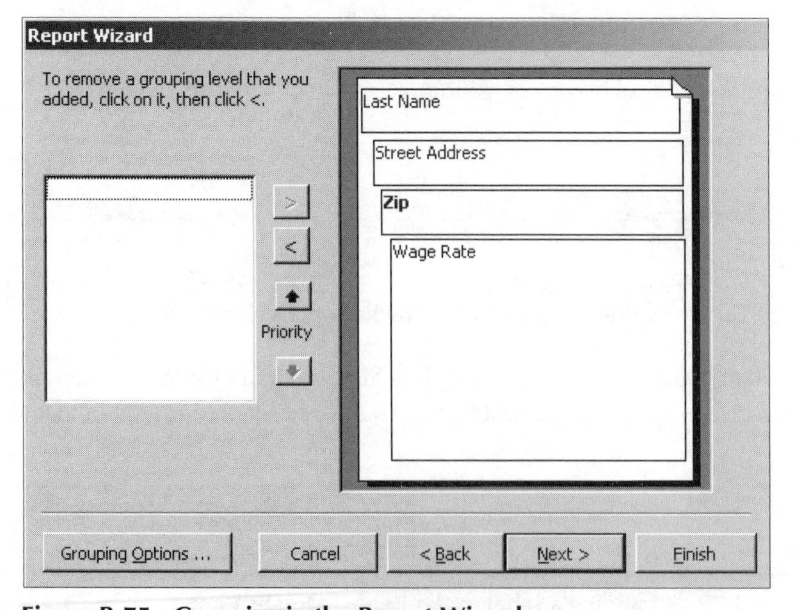

Figure B-75 Grouping in the Report Wizard

Then, follow these steps:

1. Click Next.
2. You do not Sum anything in Summary Options.
3. Click off the check mark on "Adjust the field width so all fields fit on a page".
4. Select Landscape.
5. Select Stepped. Click Next.
6. Select Corporate. Click Next.
7. Type a title (Wage Rates for Employees). Click Finish.

When you run the report, it will have a "staircase" grouped organization. In the report that follows in Figure B-76, notice that Zip data is shown below Street Address data, and Street Address data is shown below Last Name data. (The field Wage Rate is shown subordinate to all others, as desired. Wage rates may not show on the screen without scrolling).

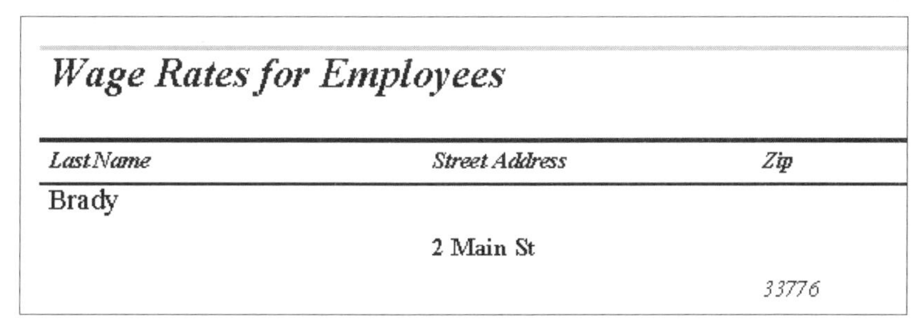

Figure B-76 Wage Rates for Employees grouped report (Wage Rate not shown)

Suppose that you want the last name, street address, and zip all on the same line. The way to do that is to take the report into the Design View for editing. At the Database window, select "Wage Rates for Employees" Report and Design. At this point, the headers look like those shown in Figure B-77.

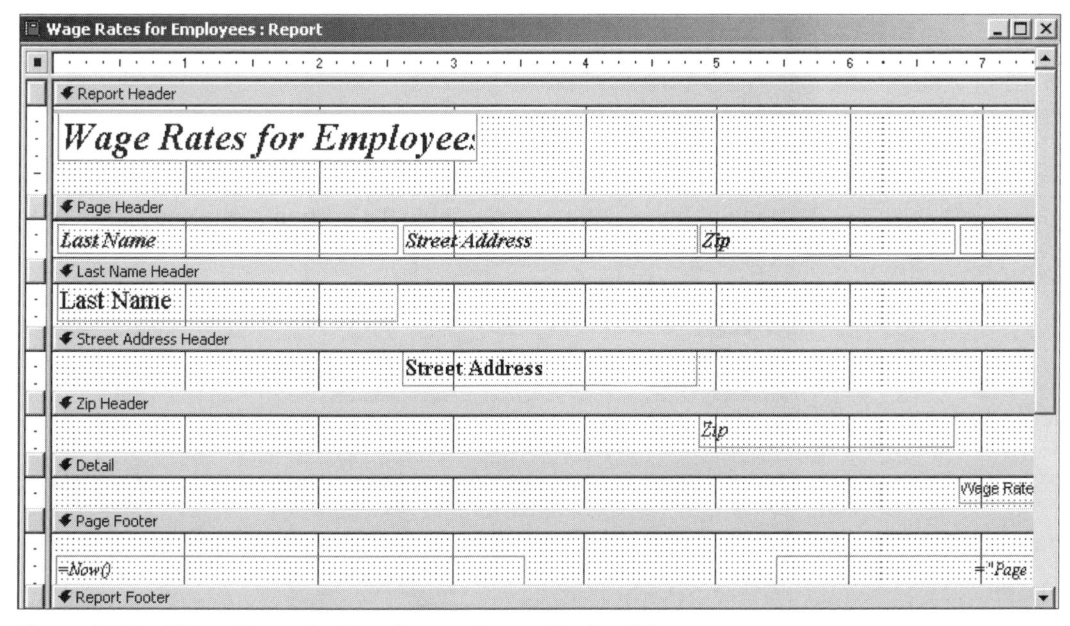

Figure B-77 Wage Rates for Employees report Design View

Your goal is to get the Street Address and Zip fields into the last name header (*not* into the page header!), so they will then print on the same line. The first step is to click the Street Address object in the Street Address Header, as shown in Figure B-78.

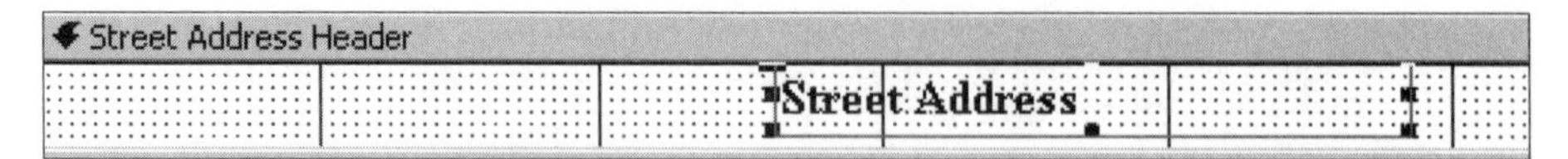

Figure B-78 Selecting Street Address object in the Street Address header

Hold down the button with the little hand icon, and drag the object up into the Last Name Header, as shown in Figure B-79.

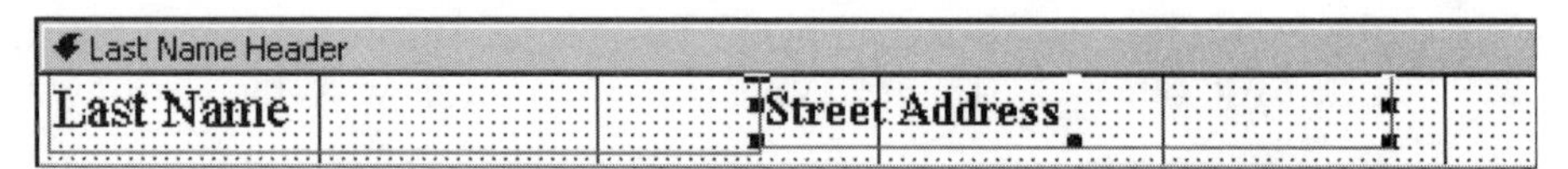

Figure B-79 Moving the Street Address object to the Last Name header

Do the same thing with the Zip object, as shown in Figure B-80.

Figure B-80 Moving the Zip object to the Last Name header

To get rid of the header space allocated to the objects, tighten up the "dotted" area between each header. Put the cursor on the top of the header panel. The arrow changes to something that looks like a crossbar. Click and drag it up to close the distance. After both headers are moved up, your screen should look like that shown in Figure B-81.

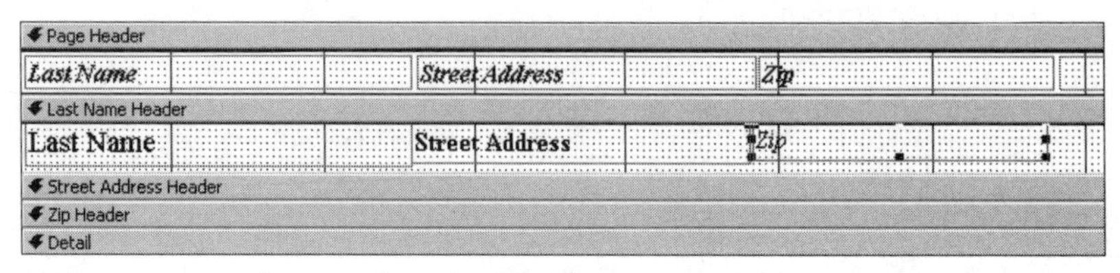

Figure B-81 Adjusting header space

Your report should now resemble the portion of the one shown in Figure B-82.

Wage Rates for Employees

Last Name	*Street Address*	*Zip*
Brady	2 Main St	33776
Howard	28 Sally Dr	19702
Jones	18 Spruce St	19716

Figure B-82 Wage Rates for Employees report

Importing Data

Text or spreadsheet data is easily imported into Access. In business, importing data happens frequently due to disparate systems. Assume that your healthcare coverage data is on the Human Resources Manager's computer in an Excel spreadsheet. Open the software application Microsoft Excel. Create that spreadsheet in Excel now, using the data shown in Figure B-83.

	A	B	C
1	SSN	Provider	Level
2	114-11-2333	BlueCross	family
3	123-45-6789	BlueCross	family
4	148-90-1234	Coventry	spouse
5	222-82-1122	None	none
6	714-60-1927	Coventry	single
7	Your SSN	BlueCross	single

Figure B-83 Excel data

Save the file, then close it. Now you can easily import that spreadsheet data into a new table in Access. With your **Employee** database open and Tables object selected, click New and click Import Table, as shown in Figure B-84. Click OK.

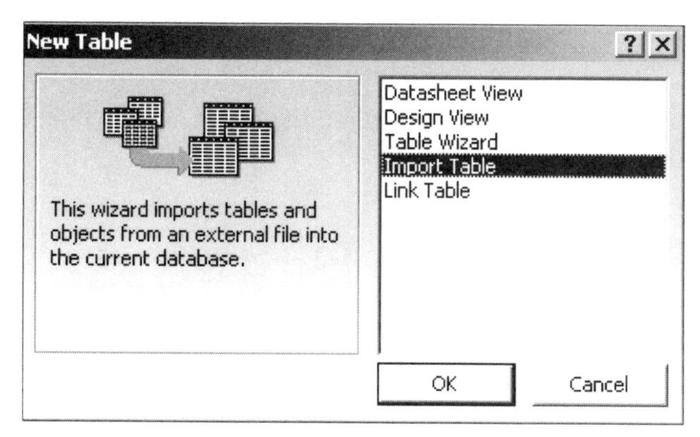

Figure B-84 Importing data into a new table

Find and import your spreadsheet. Be sure to choose **Microsoft Excel** as **Files of Type**. Assuming that you just have one worksheet in your Excel file, your next screen looks like that shown in Figure B-85.

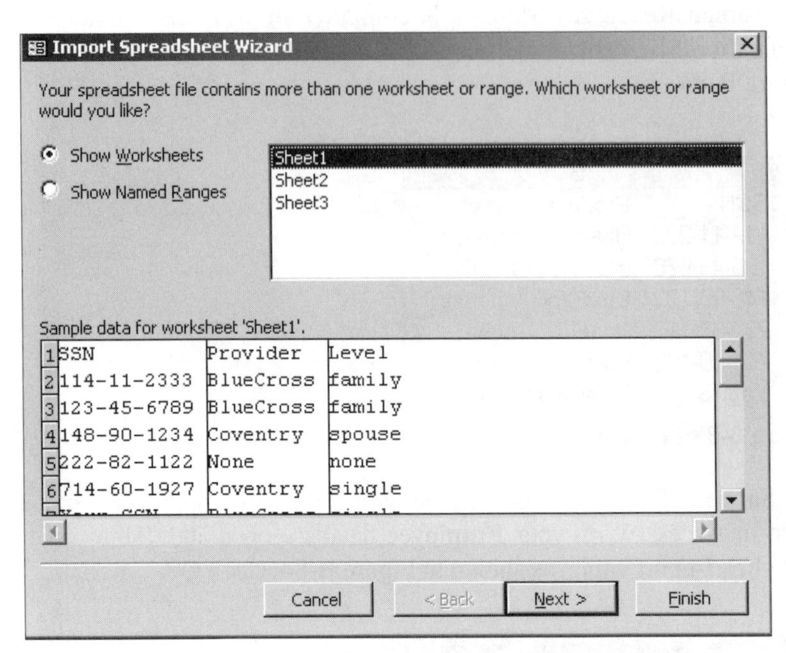

Figure B-85 First screen in the Import Spreadsheet Wizard

Choose Next, and then make sure you click the box that says First Row Contains Column Headings, as shown in Figure B-86.

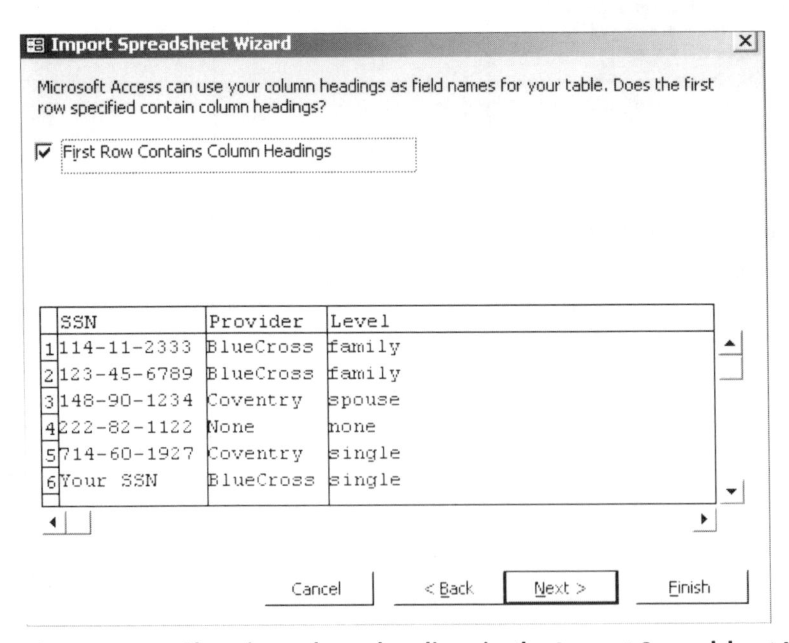

Figure B-86 Choosing column headings in the Import Spreadsheet Wizard

Store your data in a new table, do not index anything (next two screens of the Import Wizard), but choose your own primary key, which would be SSN, as chosen in Figure B-87.

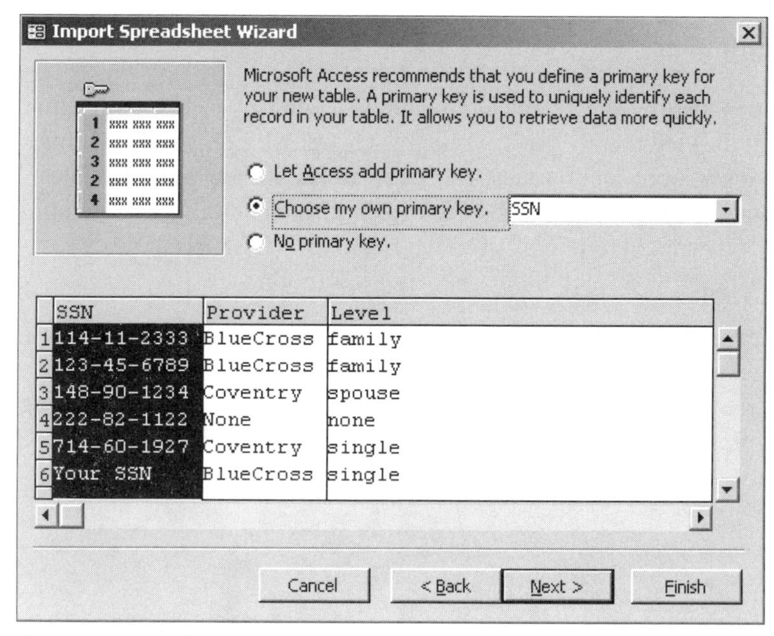

Figure B-87 Choosing a primary key field in the Import Spreadsheet Wizard

Continue through the Wizard, giving your table an appropriate name. After the table is imported, take a look at it and its design. (Highlight Table, Design button.) Note the width of each field (very large). Adjust the field properties as needed.

FORMS

Forms simplify adding new records to a table. The Form Wizard is easy to use and can be performed on a single table or on multiple tables.

When you base a form on one table, you simply identify that table when you are in the Form Wizard set-up. The form will have all the fields from that table and only those fields. When data is entered into the form, a complete new record is automatically added to the table.

But what if you need a form that includes the data from two (or more) tables? Begin (counterintuitively) with a query. Bring all tables you need in the form into the query. Bring down the fields you need from each table. (For data entry purposes, this probably means bringing down *all* the fields from each table.) All you are doing is selecting fields that you want to show up in the form, so you make *no criteria* after bringing fields down in the query. Save the query. When making the form, tell Access to base the form on the query. The form will show all the fields in the query; thus, you can enter data into all the tables at once.

Suppose that you want to make one form that would, at the same time, enter records into the EMPLOYEE table and the WAGE DATA table. The first table holds relatively permanent data about an employee. The second table holds data about the employee's starting wage rate, which will probably change.

The first step is to make a query based on both tables. Bring all the fields from both tables down into the lower area. Basically the query just gathers up all the fields from both tables into one place. No criteria are needed. Save the query.

The second step is to make a form based on the query. This works because the query knows about all the fields. Tell the form to display all fields in the query. (Common fields—here, SSN—would appear twice, once for each table.)

Forms with Subforms

You can also make a form that contains a subform. This application would be particularly handy for viewing all hours worked each week by employee. Before you create a form that contains a subform, you must form a relationship between the tables. Suppose that you want to show all the fields from the EMPLOYEE table, and for each employee, you want to show the hours worked (all fields from the HOURS WORKED table).

Join the Tables

To begin, first form a relationship between those two tables by joining them: Choose the Tables object and then choose Tools-Relationships. The Show Table dialog box will pop up. Add the EMPLOYEE table and the HOURS WORKED table. Drag your cursor from the SSN field in the EMPLOYEE table to the SSN field in the HOURS WORKED table. Another dialog box will pop up, as shown in Figure B-88.

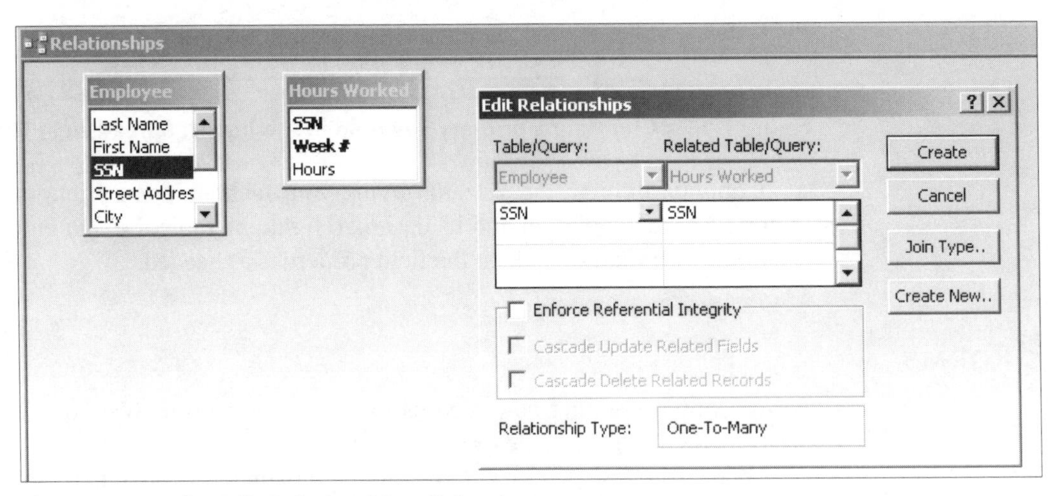

Figure B-88 The Edit Relationships dialog box

Click the Join Type button, and choose Number 2: *Include ALL records from 'Employee' and only those records from 'Hours Worked' where the joined fields are equal*, as shown in Figure B-89:

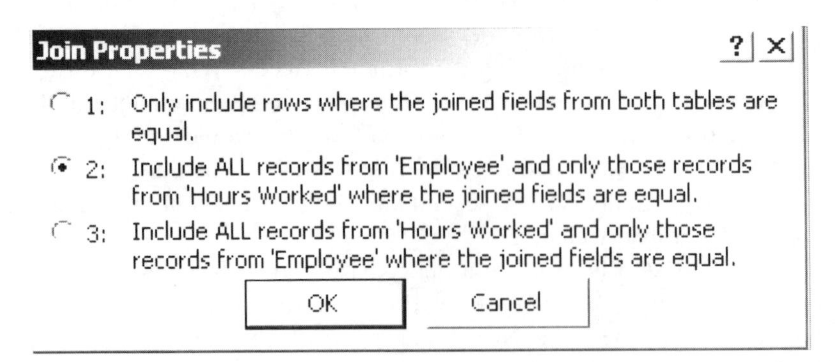

Figure B-89 The Join Properties dialog box

Click OK, then click Create. Close the Edit Relationships window and save the changes.

Create the Form and Subform

To create the form and subform, first create a simple, one-table form using the Form Wizard on the EMPLOYEE table. Follow these steps:

1. In the Forms Object, choose Create form **by using Wizard**.
2. Make sure the table Employee is selected under the drop-down menu of Tables/Queries.
3. Select all Available Fields by clicking the right double-arrow button.
4. Select Next.
5. Select Columnar layout.
6. Select Next.
7. Select Standard Style.
8. Select Next.
9. When asked, "What title do you want for your form?," type Employee Hours.
10. Select Finish.

After the form is complete, click on the Design View, so your screen looks like the one shown in Figure B-90.

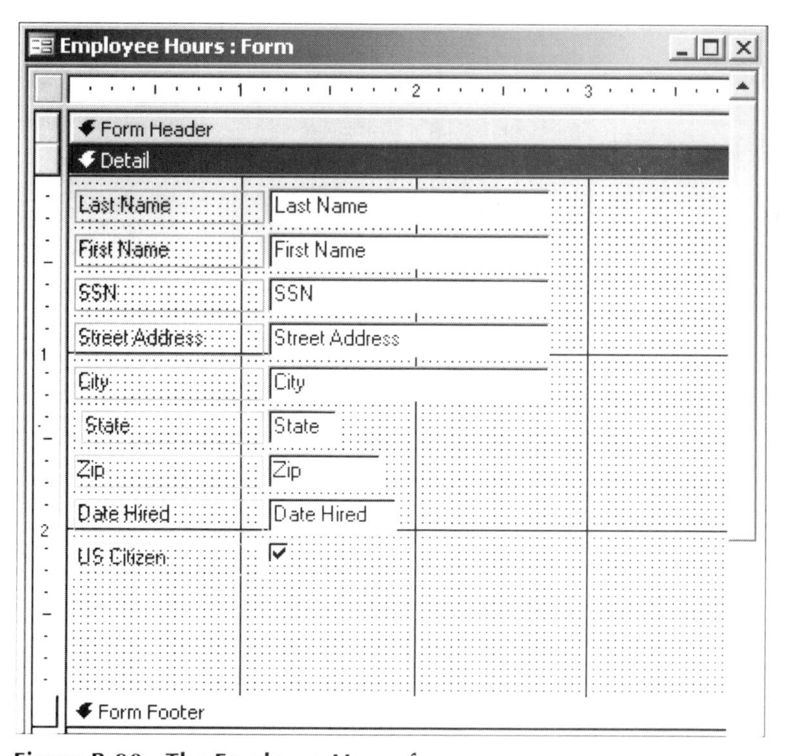

Figure B-90 The Employee Hours form

Make sure the Toolbox window is showing on the screen (Figure B-91). If it is not visible, select View—Toolbox (The Toolbox may also appear as a toolbar for some students.)

Figure B-91 The Toolbox window

Click the Subform/Subreport button (6th row, button on right) and, using your cursor, drag a small section next to the State, Zip, Date Hired, and US Citizen fields in your form design. As you lift your cursor, the Subform Wizard will appear, as shown in Figure B-92.

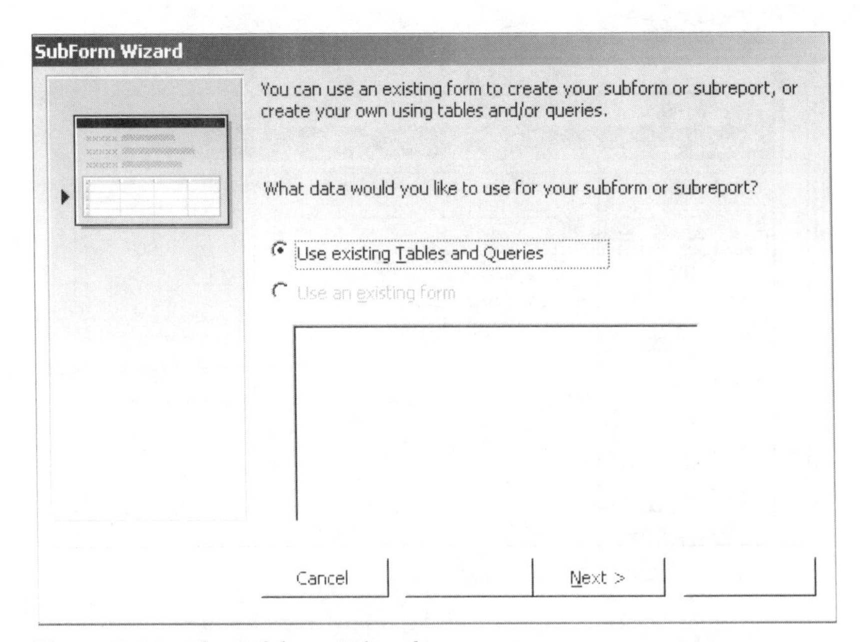

Figure B-92 The Subform Wizard

Follow these steps to create data in the subform.

1. Select the button Use Existing Tables and Queries.
2. Select Next.
3. Under Tables/Queries, choose the HOURS WORKED table, and bring all fields into the Selected Fields box by clicking the right double-arrow button.
4. Select Next.

5. Select the Choose **from a list** radio button.
6. Select Next.
7. Use the default subform name.
8. Select Finish.

Now you will need to adjust the design so all fields' data are visible. Go to the Datasheet view, and click through the various records to see how the subform data changes. Your final form should resemble the one shown in Figure B-93.

Tutorial B

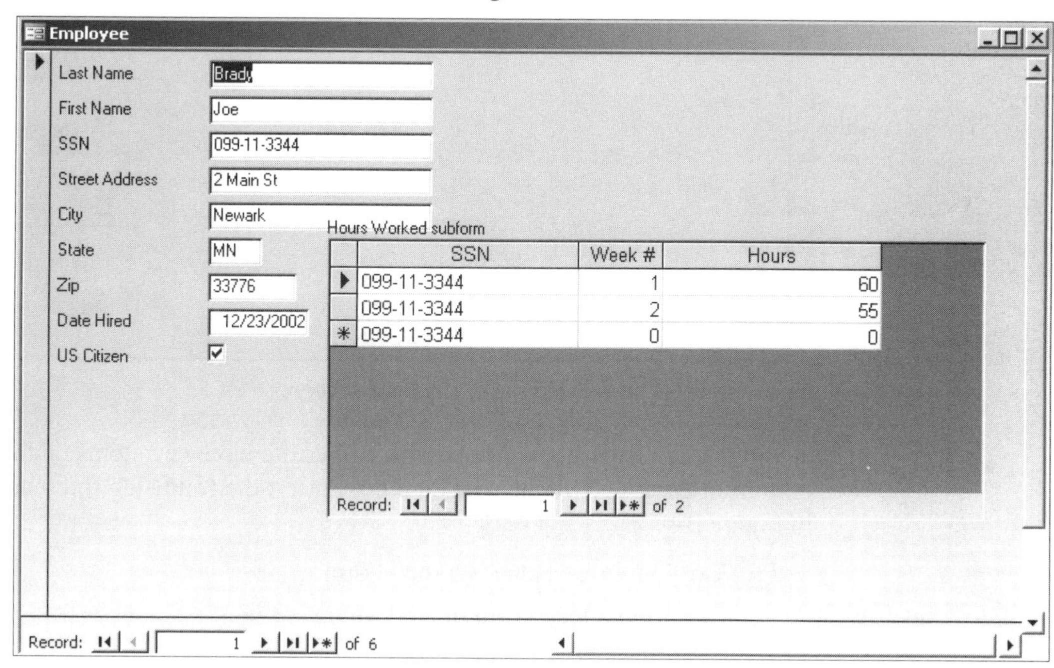

Figure B-93 The Employee Hours form with the Hours Worked subform

Create a Switchboard Form

If you want someone who knows nothing about Access to run your Access database, you can use the Switchboard Manager to create a Switchboard form to simplify their work. A Switchboard form provides a simple, user-friendly interface that has buttons to click to do certain tasks. For example, you could design a Switchboard with three buttons: one for the Employee Hours Worked form, one for the Wage Rates for Employees report, and one for the Hours Worked by Week report. Your finished product will be a page showing three buttons. Each button can be clicked to open either the form, or one of the two reports. To design that Switchboard, use the following steps.

1. Remain on the Forms Object.
2. Select Tools.
3. Select Database Utilities.
4. Select Switchboard Manager.
5. A screen will prompt you with the question, "The Switchboard Manager was unable to find a valid switchboard in this database. Would you like to create one?" Click Yes.

The Switchboard Manager screen will open, as shown in Figure B-94. Leaving the Switchboard (Default) highlighted, click the Edit button.

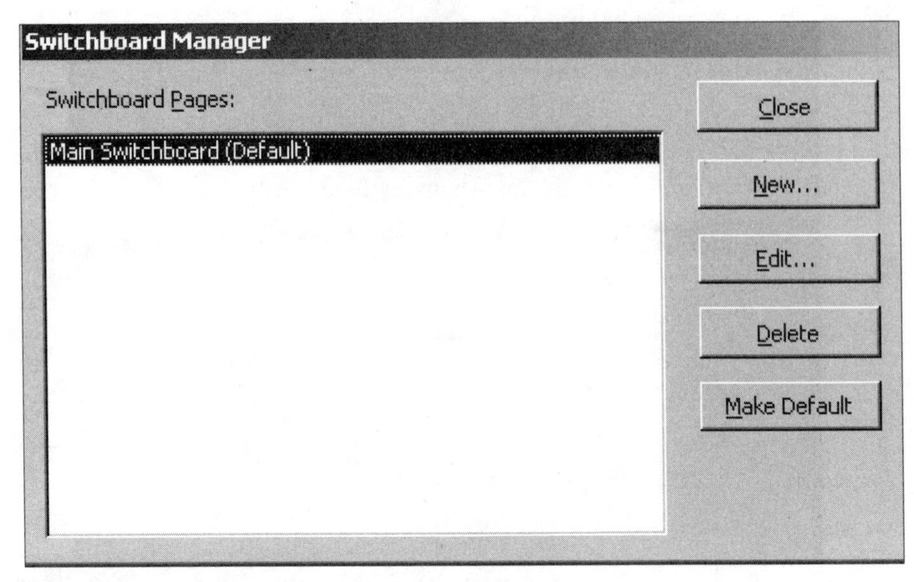

Figure B-94 The Switchboard Manager screen

In the Edit Switchboard page, you will create three new items on the page. Click the New button. In the Edit Switchboard Item box, insert the following three items of data (as shown in Figure B-95):

1. *Text:* Employee Hours Worked Form
2. *Command:* Open Form in Add Mode
3. *Form:* Employee Hours

Click OK when you are finished.

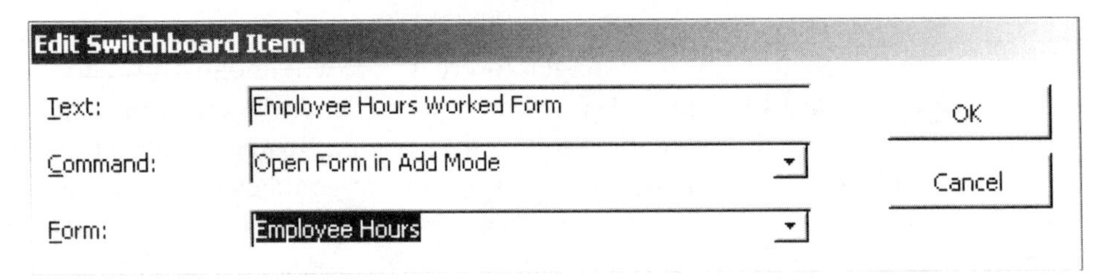

Figure B-95 The Edit Switchboard Item screen

You will repeat this procedure two more times (i.e., Click the New button in the Edit Switchboard Page). Next, insert the following data:

1. *Text:* Wage Rate for Employees Report
2. *Command:* Open Report
3. *Report:* Wage Rate for Employees

Click OK when you are finished. Then, repeat the procedure (i.e. Click the New button in the Edit Switchboard Page) and insert the following data:

1. *Text:* Hours Worked by Week Report

2. *Command:* Open Report
3. *Report:* Hours Worked by Week

Click OK when you are finished. At this point, your Edit Switchboard screen should look like Figure B-96.

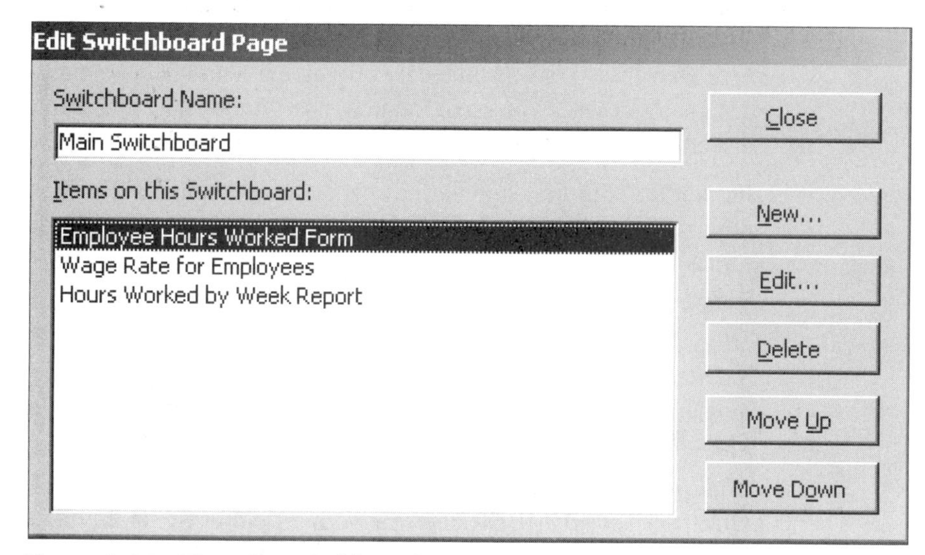

Figure B-96 The Edit Switchboard Page

Click the Close button, and then click the Close button again.

You can test the Switchboard by clicking the Switchboard in the Forms Objects. It should look like that shown in Figure B-97.

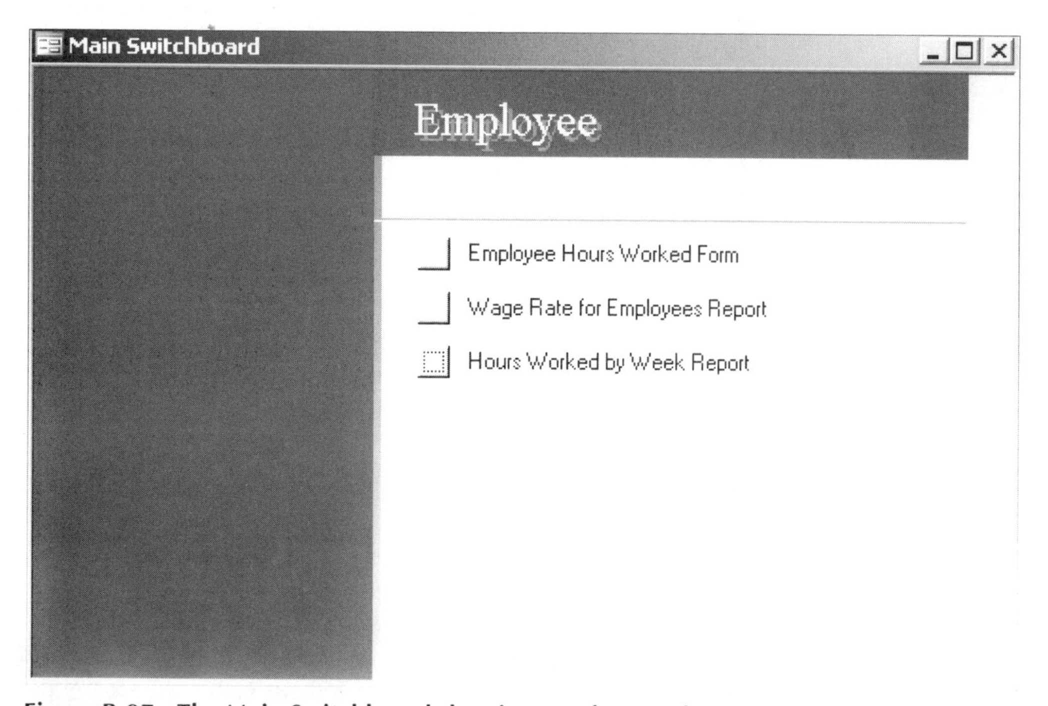

Figure B-97 The Main Switchboard showing one form and two reports

Troubleshooting Common Problems

Access beginners (and veterans!) sometimes create databases that have problems. Common problems are described here, along with their causes and corrections.

1. *"I saved my database file, but it is not on my diskette! Where is it?"*

   You saved to some fixed disk. Use the Search option of the Windows Start button. Search for all files ending in ".mdb" (search for *.mdb). If you did save it, it is on the hard **Drive (C:\)** or on some network drive. (Your site assistant can tell you the drive designators.) Once you have found it, use Windows Explorer to copy it to your diskette in **Drive A:**. Click it, and drag to **Drive A:**.

   Reminder: Your first step with a new database should be to Open it on the intended drive, which is usually **Drive A:** for a student. Don't rush this step. Get it right. Then, for each object made, save it *within* the current database file.

2. *"What is a 'duplicate key field value'? I'm trying to enter records into my Sales table. The first record was for a sale of product X to customer #101, and I was able to enter that one. But when I try to enter a second sale for customer #101, Access tells me I already have a record with that key field value. Am I only allowed to enter one sale per customer!?"*

   Your primary key field needs work. You may need a compound primary key—CUSTOMER NUMBER and some other field or fields. In this case, CUSTOMER NUMBER, PRODUCT NUMBER, and DATE OF SALE might provide a unique combination of values—or consider using an INVOICE NUMBER field as a key.

3. *"My query says 'Enter Parameter Value' when I run it. What is that?"*

   This symptom, 99 times out of 100, indicates you have an expression in a Criteria or a Calculated Field, and *you misspelled a field name in the expression*. Access is very fussy about spelling. For example, Access is case sensitive. Furthermore, if you put a space in a field name when you define the table, then you must put a space in the field name when you reference it in a query expression. Fix the typo in the query expression.

   This symptom infrequently appears when you have a calculated field in a query, and you elect *not* to show the value of the calculated field in the query output. (You clicked off the Show box for the calculated field.) To get around this problem, click Show back on.

4. *"I'm getting a fantastic number of rows in my query output—many times more than I need. Most of the rows are duplicates!"*

   This symptom is usually caused by a failure to link together all tables you brought into the top half of the query generator. The solution is to use the manual click-and-drag method. Link the fields (usually primary key fields) with common *values* between tables. (Spelling of the field names is irrelevant because the link fields need not be spelled the same.)

5. *"For the most part, my query output is what I expected, but I am getting one or two duplicate rows."*

   You may have linked too many fields between tables. Usually only a single link is needed between two tables. It's unnecessary to link each common field in all combinations of tables; usually it's enough to link the primary keys. A layman's explanation for why over-linking causes problems is that excess linking causes Access to "overthink" the problem and repeat itself in its answer.

On the other hand, you might be using too many tables in the query design. For example, you brought in a table, linked it on a common field with some other table, but then did not use the table. You brought down none of its fields and/or you used none of its fields in query expressions. Therefore, get rid of the table, and the query should still work. Try doing this to see whether the few duplicate rows disappear: Click the unneeded table's header in the top of the QBE area and press the Delete key.

6. *"I expected six rows in my query output, but I only got five. What happened to the other one?"*

   Usually this indicates a data-entry error in your tables. When you link together the proper tables and fields to make the query, remember that the linking operation joins records from the tables *on common values* (*equal* values in the two tables). For example, if a primary key in one table has the value "123", the primary key or the linking field in the other table should be the same to allow linking. Note that the text string "123" is not the same as the text string "123 " —the space in the second string is considered a character too! Access does not see unequal values as an error: Access moves on to consider the rest of the records in the table for linking. Solution: Look at the values entered into the linked fields in each table and fix any data entry errors.

7. *"I linked fields correctly in a query, but I'm getting the empty set in the output. All I get are the field name headings!"*

   You probably have zero common (equal) values in the linked fields. For example, suppose you are linking on Part Number (which you declared as text): in one field you have part numbers "001", "002", and "003", and in the other table part numbers "0001", "0002", and "0003". Your tables have no common values, which means no records are selected for output. You'll have to change the values in one of the tables.

8. *"I'm trying to count the number of today's sales orders. A Sigma query is called for. Sales are denoted by an invoice number, and I made this a text field in the table design. However, when I ask the Sigma query to 'Sum' the number of invoice numbers, Access tells me I cannot add them up! What is the problem?"*

   Text variables are words! You cannot add words, but you can count them. Use the Count Sigma operator (not the Sum operator): count the number of sales, each being denoted by an invoice number.

9. *"I'm doing Time arithmetic in a calculated field expression. I subtracted the Time In from the Time Out and I got a decimal number! I expected 8 hours, and I got the number .33333. Why?"*

   [Time Out] – [Time In] yields the decimal percentage of a 24-hour day. In your case, 8 hours is a third of a day. You must complete the expression by multiplying by 24: ([Time Out] – [Time In]) * 24. Don't forget the parentheses!

10. *"I formatted a calculated field for currency in the query generator, and the values did show as currency in the query output; however, the report based on the query output does not show the dollar sign in its output. What happened?"*

    Go into the report Design View. There is a box in one of the panels representing the calculated field's value. Click the box and drag to widen it. That should give Access enough room to show the dollar sign, as well as the number, in output.

11. *"I told the Report Wizard to fit all my output to one page. It does print to just one page. But some of the data is missing! What happened?"*

Tutorial B

Access fits the output all on one page by *leaving data out*! If you can stand to see the output on more than one page, click off the "Fit to a Page" option in the Wizard. One way to tighten output is to go into the Design View and remove space from each of the boxes representing output values and labels. Access usually provides more space than needed.

12. *"I grouped on three fields in the Report Wizard, and the Wizard prints the output in a staircase fashion. I want the grouping fields to be on one line! How can I do that?"*

    Make adjustments in the Design View. See the Reports section of this tutorial for instruction.

13. *"When I create an Update query, Access tells me that zero rows are updating, or more rows are updating than I want. What is wrong?"*

    If your Update query is not correctly set up, for example, if the tables are not joined properly, it will either try not to update anything, or it will update all the records. Check the query, make corrections, and run it again.

14. *"After making a Summation Query with a Sum in the Group By row and saving that query, when I go back to it, the Sum field now says Expression, and Sum is put in the field name box. Is this wrong?"*

    Access sometimes changes that particular statistic when the query is saved. The data remains the same, and you can be assured your query is correct.

The Misty Valley Country Inn

1
CASE

Setting Up a Relational Database to Create Tables, Queries, and Reports

Preview

In this case, you'll create a relational database for a country inn. First, you'll create three tables and populate them with data. Next, you'll create two queries: a Summation query and a Parameter query. Finally, you'll create a report that shows the revenue received from customers.

Preparation

- Before attempting this case, you should have experience using Microsoft Access.
- Complete any part of the previous tutorials that your instructor assigns, or refer to them as necessary.
- In this case, you will be using the following features of Access: Select queries, Summation queries, and Grouped reports.

BACKGROUND

Fran and Stella Sardino own and run a small bed and breakfast, called the Misty Valley Country Inn, in the Pocono Mountains of eastern Pennsylvania. They attract repeat customers during two busy periods of the year: In winter, customers are attracted by holiday shopping at the local factory outlets and snow skiing; in spring, customers are attracted by the local golf course. The remainder of the year, their seven-unit inn remains fairly empty. After some researching on the Internet, Fran discovers that the Misty Valley Country Inn could join a national chain called Bed and Breakfast Inns of America (BBIA). By joining this chain, their inn would have access to a wider variety of marketing outlets to generate more business.

Fran and Stella contact BBIA, and the president is excited about adding another inn to the group; however, he requires the Sardino sisters to computerize their reservation system. Not knowing anything about computers or databases, the Sardinos contact your college and hire you as a consultant to create the reservation system.

According to the computer database specifications of the BBIA, each inn is required to have three tables in their reservation database. One table, the ROOMS table, contains data about the configuration of the room (number of double beds, number of single beds), where it is located in the inn (which floor), whether it is "on-suite" (meaning it has its own private bathroom), and the cost per night. Another table, the CUSTOMERS table, keeps track of the particulars of each customer, including their name, address, telephone number, and whether they bring their pets. Pets are allowed at the inn if an extra deposit is submitted when checking in. Some guests like to bring their pets during the spring cat and dog shows. The third and final table, RESERVATIONS, provides each customer with a customer ID number and records the room number reserved for each customer on specific dates. Additional information, such as deposits, will continue to be kept by hand until the owners are comfortable with the computer system.

The Sardino sisters want to use this database to determine which rooms are generating the most income. If they find that a certain type of room is much more popular than another, they may reconfigure their B&B to fit their customers' needs. So, the first query will sum the amount of money generated by each room, in order of the most money generated to the least.

In addition, there are times when the owners are very busy and need to quickly answer questions about room availability. They will need another query to list the rooms booked for a certain week.

Finally, because the sisters have so many repeat customers, they would like to cater to their best clientele. To target these customers, they need a report that lists each customer, the nights stayed, and the amount of revenue that each customer generates.

ASSIGNMENT 1 CREATING TABLES

Use Microsoft Access to create the tables with the fields shown in Figures 1-1 through 1-3 and discussed in the Background section. Populate the database tables as shown. Add your name to the CUSTOMERS table as Customer Number 110, replacing Luke Pickett, leaving the other fields in that record as they are. To minimize typing, you will only record reservations for December 2005. Save your database as **Misty Valley Country Inn.mdb**.

Case 1

Customers : Table

Customer Number	Last Name	First Name	Street	City	State	Zip	Telephone	Pets
101	Fontanella	Amy	234 Maple Lane	Newark	DE	19711	(302)737-0090	☐
102	Mellow	Robert	56 East HWY I	Wilmington	DE	19808	(302)738-7623	☑
103	Holmes	Martha	10 Danbury Circle	Bear	DE	19756	(302)821-0091	☐
104	O'Brien	Sue	1500 West New London	Newark	DE	19711	(302)737-5543	☑
105	Hall	Michael	5 Oak Avenue	Paramus	NJ	07664	(201)992-1299	☐
106	Sanchez	Anita	89 Wayne Dr	Richfield	NJ	07883	(201)876-0098	☐
107	Cobb	Sylvester	90 Rt 273	Elkton	MD	21921	(410)398-7654	☑
108	Golden	Laurie	8900 Highrise Circle	Northeast	MD	21931	(410)895-7643	☐
109	Patel	Sammie	76 Bell Court	Landenberg	PA	20192	(610)554-6512	☑
110	Pickett	Luke	90 South Windsor	Landenberg	PA	20192	(610)554-9021	☐

Figure 1-1 The CUSTOMERS table

Reservations : Table

Room Number	Customer Number	Check-in Date	Check-out Date
1	108	12/1/2005	12/5/2005
1	110	12/13/2005	12/14/2005
2	101	12/15/2005	12/17/2005
2	110	12/23/2005	12/26/2005
3	103	12/1/2005	12/5/2005
3	104	12/7/2005	12/9/2005
3	110	12/20/2005	12/28/2005
4	102	12/2/2005	12/3/2005
4	107	12/5/2005	12/7/2005
4	107	12/15/2005	12/17/2005
6	106	12/18/2005	12/19/2005
7	105	12/15/2005	12/24/2005

Figure 1-2 The RESERVATIONS table

Rooms : Table

Room Number	Floor	Number of Double Beds	Number of Single Beds	Bath	Price Per Night
1	1	1	0	☐	$75.00
2	1	1	0	☐	$75.00
3	2	1	2	☑	$120.00
4	3	2	0	☑	$140.00
5	3	1	1	☑	$110.00
6	4	0	1	☐	$60.00
7	4	0	1	☐	$60.00

Figure 1-3 The ROOMS table

Assignment 2 Creating Queries and Reports

Assignment 2A: Creating a Summation Query

Create a Summation query that will calculate the amount of money generated by each room's rental. Beginning with a Select query, you first need to create a calculated field that figures the money generated by each room. Call that calculated field Total. Now, using the Summation function, aggregate the amount of money brought in for each room. Sort Descending on Total. Making sure column headings look professional. Your output should resemble that shown in Figure 1-4.

Money Generated by Room : Select Query

Room Number	Total
3	$1,680.00
4	$700.00
7	$540.00
2	$375.00
1	$375.00
6	$60.00

Figure 1-4 Query: Money Generated by Room

Save the query as Money Generated by Room and print the results.

Assignment 2B: Creating a Select Query

Create a query that lists all the reservations for the week of December 12, 2005. The output should be Room Number, Floor, Check-in Date, and Check-out Date. Your output should resemble that in Figure 1-5.

Rooms Booked the Week of 12/12/05 : Select Query

Room Number	Floor	Check-in Date	Check-out Date
2	1	12/15/2005	12/17/2005
7	4	12/15/2005	12/24/2005
1	1	12/13/2005	12/14/2005
4	3	12/15/2005	12/17/2005
6	4	12/18/2005	12/19/2005

Figure 1-5 Query: Rooms Booked the Week of 12/12/05

Save the query as Rooms Booked the Week of 12/12/05. Print the output.

Assignment 2C: Generating the Revenue Report

Generate a report that groups each customer and shows the dates they stayed at the B&B and how much money they spent renting a room. To create the report, you will need to do the following:

- You must first create the query that will bring all the data together for the report, including a calculated field to figure out each customer's bill for his or her stay at the B&B.
- In the Report Wizard, bring in the following fields: Last Name, First Name, Check-in Date, and Bill.
- Group by Last Name and First Name.
- In the Report Wizard, title the report Revenue From Customers—December 2005.
- You will find that the First Name field needs to be moved to the same line as the Last Name in the Design View.
- Use Print Preview to make sure the report will print correctly.

- You will have to adjust the report design in the Design View so the report resembles the portion of the report shown in Figure 1-6 (with other adjustments to make the report look correct).
- Save the report as Revenue From Customers—December 2005.

Revenue From Customers - December 2005

LastName	*FirstName*	*Check-in Date*	*Bill*
Cobb	*Sylvester*		
		12/15/2005	$280.00
		12/5/2005	$280.00
	Total Revenue per customer		*$560.00*
Fontanella	*Amy*		
		12/15/2005	$150.00
	Total Revenue per customer		*$150.00*

Figure 1-6 Report: Revenue From Customers—December 2005

Make sure you close the database file before removing your disk.

DELIVERABLES

1. Printouts of three tables
2. Query output: Money Generated by Room
3. Query output: Rooms Booked the Week of 12/12/05
4. Report: Revenue From Customers—December 2005
5. Disk or CD with database file

Staple all pages together. Put your name and class number at the top of the page. Make sure your disk or CD is labeled.

2

CASE

The Outsourcing Bookkeeping and Payroll Company

DESIGNING A RELATIONAL DATABASE TO CREATE TABLES, QUERIES, AND REPORTS

PREVIEW

In this case, you'll design a relational database for a company that offers bookkeeping and payroll services to other companies. After your database design is completed and correct, you will create database tables and populate them with data. Next, you'll create a form for the employees of the firm to record their hours worked for clients. Then, you'll create two queries: one to list all the accounts in a given town, and another to calculate the number of hours worked on a given day for each client. You'll finish this case by creating two reports: one to display client accounts grouped by town, and one to report the number of hours worked on each type of task.

PREPARATION

- Before attempting this case, you should have some experience in database design and in using Microsoft Access.
- Complete any part of Database Design Tutorial A that your instructor assigns.
- Complete any part of Access Tutorial B that your instructor assigns, or refer to the tutorial as necessary.
- Refer to Tutorial E as necessary.
- In this case, you will be using the following features of Access: Forms, Parameter queries, Calculated Fields in queries, and Grouped reports.

BACKGROUND

In today's business environment, outsourcing is very popular. When a company outsources a business operation, such as the company's bookkeeping or payroll, they hire another company to do that job rather than having an employee do it. Both large and small companies outsource work to save money and time—and to have a more accurate job done by a specialist in the field.

Outsourcing even a simple job, such as payroll, can result in a substantial savings. For example, a small company with 10 employees has a typical payroll cost of $2,600. It is not cost-effective to hire another in-house person to perform that job. Similarly, it might not be wise to have a non-specialist employee take on the job. The IRS claims that almost one half of all small businesses pay over $750 per year for late or incorrect filings. If an outsource specialist does payroll, such fines can be avoided because the outsource specialist will have the latest version of the tax tables and government forms. In addition, outsourcer specialists can also deal with bookkeeping and other payroll-related items, such as electronic tax payments, employee direct deposits, custom reports, and much more.

For generations, a company called Outsourcing Bookkeeping and Payroll (OBP) has helped small businesses manage their accounts. You have landed a summer internship at OBP, based on your database experience. The management wants you to design a database for them to keep track of their clients, employees, and the outsourced jobs they do.

First, they need to get their client records in order. Currently, all client data is recorded on an old-fashioned Rolodex file system. Because most clients prefer e-mail communication, the Rolodex is archaic. (Using a Rolodex *was* a popular way for businesses to keep track of clients. Each client's information was listed on a separate card; all cards were held together by two rings. To find a client, you could spin the side wheels of the Rolodex and "roll" through the various clients.) OBP needs to keep track of the clients, their addresses, contact person, telephone, and now increasingly, e-mail address. Some clients have similar names but are located in different cities.

In addition to maintaining employee information, such as Social Security Number and birth date, the company wants to allow their employees to submit their client work hours into the database directly, via a form. From there, the management of OBP can generate a report to determine which clients are using which employees, and for how long.

Some of the OBP salespeople have heard about your internship and have requested that you create a query that they can run to identify which clients are in which towns. When OBP salespeople are on the road, it's more efficient for them to call on all their clients in a given area. With the Rolodex system, it's very difficult to find that information because data is arranged alphabetically by client name. In addition, having a report that groups clients by town would allow management to have a better grasp of their market demographics.

ASSIGNMENT 1 CREATING THE DATABASE DESIGN

In this assignment, you will design your database tables on paper, using a word-processing program. Pay close attention to the tables' logic and structure. Do not start your Access code (Assignment 2) before getting feedback from your instructor on Assignment 1. Keep in mind that you will need to look at what is required in Assignment 2 to design your fields and tables properly. It's good programming practice to look at the required outputs before designing your database. When designing the database, observe the following guidelines:

- First, determine the tables you'll need by listing on paper the name of each table and the fields that it should contain. Avoid data redundancy. Do not create a field if it could be created by a "calculated field" in a query.
- You'll need a transaction table. Avoid duplicating data.
- Document your tables by using the Table facility of your word processor. Your word-processed tables should resemble the format of the table in Figure 2-1.
- You must mark the appropriate key field(s). You can designate a key field by placing an asterisk (*) next to the field name. Keep in mind that some tables need a compound primary key to uniquely identify a record within a table.
- Print out the database design.

TABLE NAME	
Field Name	*Data Type (text, numeric, currency, etc.)*
...	...
...	...

Figure 2-1 Table design

Have your design approved before beginning Assignment 2; otherwise, you may need to redo Assignment 2.

ASSIGNMENT 2 CREATING A DATABASE AND DEVELOPING A FORM, QUERIES, AND REPORTS

In this assignment, you will first create database tables in Access and populate them with data. Name and save your database as **OUTSOURCING.mdb**. Next, you will create one form, two queries, and two reports.

Assignment 2A: Creating Tables in Access

In this part of the assignment, you will create your tables in Access. Observe the following guidelines:

- Type records into the tables, using the Clients and their addresses shown in Figure 2-2. Add your name as a client, with your address and phone number. Create any additional fields that would be needed and populate them with data.
- Create employee data, including Social Security numbers and birthdates. There should be at least six employees.
- Create work dates and hours for each employee and for each client. Choose one week to record those hours and record at least 10 work sessions.
- Appropriately limit the size of the text fields; for example, a Zip field need only be 5 characters wide, not the default 50 characters in length.
- Print all tables.

Name	Address	Town	State	Zip Code	Telephone
Sandy's Snaks	23 Eastern Avenue	New Brunswick	NJ	08902	(201)854-1177
Papa's Pizza	165 Western Way	Piscataway	NJ	08903	(201)993-1928
Home Nursing	1565 International Way	Redbank	NJ	08992	(609)990-8763
5 and 10	5 Main Street	Rahway	NJ	08911	(201)675-4637
Simon and Sons	1564 Eastern Avenue	New Brunswick	NJ	08902	(201)854-8734
The Organic Grocer	54 101 Street	NY	NY	21334	(212)765-5643
Sam's HVAC	765 103 Street	NY	NY	21334	(212)876-7899

Figure 2-2 Client data

Assignment 2B: Creating a Form, Queries, and Reports

There is one form, two queries, and two reports to generate, as outlined in the background of this case. Begin with the form.

Form: Jobs Input

Create a form that the employees can use to record their hours worked on a particular date for a particular client. Base this form on one table only (hint: the Transaction table) and use the Form Wizard. Save the form as Jobs Input. Print one record from this form. Your completed form should resemble that shown in Figure 2-3.

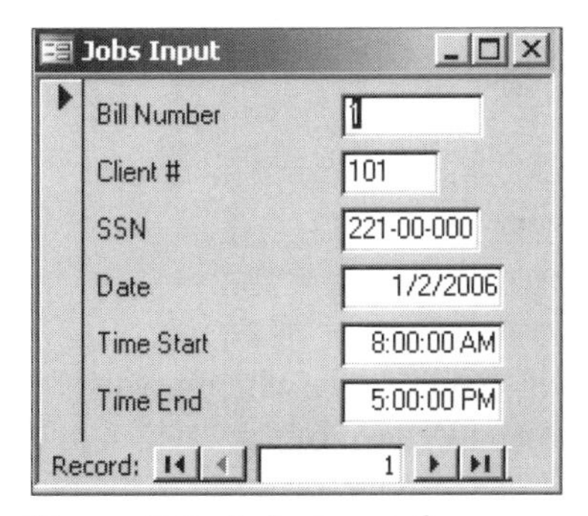

Jobs Input

Bill Number: 1
Client #: 101
SSN: 221-00-000
Date: 1/2/2006
Time Start: 8:00:00 AM
Time End: 5:00:00 PM
Record: 1

Figure 2-3 Jobs Input form

Query 1: Sales Calls

The salespeople would like to be able to run a query before they leave for a sales call to a particular town to find out which clients are in that town. Create a Parameter query that asks for the input of a town. Include in your output useful information for the salesperson such as client #, name, address, telephone, contact person, and e-mail. Your resulting query should resemble that shown in Figure 2-4, although some of your data will differ. Save the query as Sales Calls, and print the results when you input NY as the town, when prompted.

Sales Calls : Select Query

Client #	Name	Address	Telephone	Contact Person	Email
106	The Organic Grocer	54 101 Street	(212)765-5643	Lucy Shine	shine@zz.net
107	Sam's HVAC	765 103 Street	(212)876-7899	Sam Stoner	ss@comcast.net

Figure 2-4 Sales Calls query

Query 2: Hours Worked Per Client

The management of OBP would like to keep track of how many hours of employee work time each client is consuming. Create a query that calculates the total time devoted to each client. Keep in mind the following guidelines:

- In your output show only Client #, Name, and Total Time
- Total Time must be a calculated field
- Check Tutorial B for hints on time arithmetic if your numbers don't look correct
- Save the query as Hours Worked per Client

Although your data will differ, your query output should resemble that in Figure 2-5.

Hours Worked Per Client : Select Query

Client #	Name	Total Time
101	Sandy's Snaks	13
102	Papa's Pizza	3
103	Home Nursing	5
104	5 and 10	20
105	Simon and Sons	8
106	The Organic Grocer	2
107	Sam's HVAC	12

Figure 2-5 Hours Worked Per Client query

Report 1: Clients by Zip

The salespeople can now determine which clients are in which towns, but the management still does not have a feeling for how the clients break down by zip code. Create a grouped report that lists the clients by zip code. Include their name, address, state, zip, contact person, and e-mail address. Title the report Clients by Zip. Be careful to adjust the design of the report so that all fields and data are visible. You may change some column headings so that it looks tidy and readable. Your output should resemble that in Figure 2-6, although some data may differ.

Clients by Zip

Zip Code	Name	Address	Town	State	Telephone	Contact Person	Email
08902							
	Simon and Sons	1564 Eastern Avenue	New Brunswick	NJ	(201)854-8734	Fred Simon	fs@ss.com
	Sandy's Snaks	23 Eastern Avenue	New Brunswick	NJ	(201)854-1177	Amy Lee	A.Lee@ss.net
08903							
	Papa's Pizza	165 Western Way	Piscataway	NJ	(201)993-1928	Tony Maro	pap@comcast.net

Figure 2-6 Clients by Zip report

Report 2: Workers and Times for Each Client

The last report is also required by management. For each client, management would like to see the dates worked, the number of hours worked, and which employees worked those hours. In order to create this report, follow these guidelines:

- First create a query with the necessary tables.
- Calculate the time spent on each job in that query, using a calculated field.
- Use the Report Wizard based on that query.
- Group the report by the Client's Name.
- Using the Summary Options button, sum the Time.
- Remove any bogus summary bands, and make sure all data and headings are visible.

Your final report should resemble that shown in Figure 2-7, although your data should vary.

Workers and Times for Each Client

Client Name	*Employee Name*	*Date*	*Hours Worked*
5 and 10			
	Daniel Brown	1/4/2006	9
	Sandy Gomez	1/4/2006	11
Sum (in hours)			*20*
Home Nursing			
	Pierre Henri	1/2/2006	5
Sum (in hours)			5

Figure 2-7 Workers and Times for Each Client report

Assignment 3 Making a Presentation

Create a presentation for the management of OBP. Explain the database and make suggestions for future work. Include the following:

- Describe the design of your database tables.
- Tell how to use the database, enter information into the form, and run the queries and reports.
- Explain how other parts of the business might be included in an expanded version of this database project.

DELIVERABLES

1. Word-processed design of tables
2. Tables created in Access
3. Form: Jobs Input
4. Query 1: Sales Calls
5. Query 2: Hours Worked Per Client
6. Report 1: Clients by Zip
7. Report 2: Workers and Times for Each Client
8. Presentation materials
9. Any other required tutorial printouts or tutorial disk or CD

Staple all pages together. Put your name and class number at the top of the page. Make sure your disk or CD is labeled.

The Orchid Seller

3
CASE

Designing a Relational Database to Create Tables, Queries, and a Report

Preview

In this case, you'll design a relational database for an orchid seller, Bob. After your database design is completed and correct, you will create database tables and populate them with data.

Preparation

- Before attempting this case, you should have some experience in database design and in using Microsoft Access.
- Complete any part of Database Design Tutorial A that your instructor assigns.
- Complete any part of Access Tutorial B that your instructor assigns, or refer to the tutorial as necessary.
- Refer to Tutorial E as necessary.
- In this case, you will be using the following features of Access: Parameter queries, Summation queries, Select queries, Calculated Fields in queries, the Report Wizard, Grouping in a report, and Summing in a report.

Background

Last weekend when you were hanging out with a friend, her mother invited the two of you to a flower show. Even though flowers don't particularly interest you, you decided to go because there was a free dinner offered as well! At the flower show, you struck up a conversation with one of the vendors, Bob, who was selling hundreds of orchids. (The last time you saw an orchid was on a corsage at your high-school senior prom.) Bob told you that he is turning his orchid hobby into a part-time business.

Bob said that by trade he is a building contractor. A few years ago, he bought his wife a phalaenopsis orchid, also called a moth orchid. He figured that it would bloom for a week and then die, but Bob's wife put it in the kitchen window, and it bloomed for several months and didn't die when it stopped blooming.

Bob became interested in orchids, and he discovered that certain kinds of orchids make excellent houseplants. Bob bought more orchids and different kinds of orchids. Bob's orchid collection soon overflowed the kitchen, living room, and sun room. Then Bob joined an orchid society and started trading and selling orchids with others in the society. Bob built a small greenhouse for his ever-expanding collection, and then he built two even larger greenhouses. When the orchid society had an orchid show and sale, Bob rented a table and sold some of his orchids. Bob made a handsome profit.

In addition to maintaining his own collection, Bob now is buying orchids from dozens of wholesale orchid suppliers in Taiwan, Hawaii, Columbia, Florida, and California. He buys orchids in bloom and sells them at weekend orchid or flower shows. On average, Bob sells 200 orchids at one show each month. By doing this quick turnaround, Bob has found that he can make several thousand dollars for a weekend's work!

Because Bob's orchid business is growing so fast, it is becoming unmanageable. Bob needs a database both to manage his orchids and to track his profitability, and he will pay you to set up a database for him. His goals are as follows:

1. Bob needs to manage the cultural needs of orchids in his greenhouse collection and to provide the appropriate conditions for orchids that he resells. The 300 species and hybrid crosses that Bob buys (which fall into several main orchid groups) have different needs for heat, light, and humidity. After all, some of these plants come from steamy, lowland rainforests; others come from cool, cloud forests; and some grow in the shadows of the Himalaya Mountains. First, Bob needs to sort orchids by orchid group, according to their heat requirements: "warm," "intermediate," or "cool." Each of his three greenhouses is geared to one of these temperature ranges. Second, Bob positions orchids in each greenhouse according to their light needs: plants needing "high" light go on top shelves or are positioned closest to the glass walls; plants needing "intermediate" or "low" light are more toward the middle of a greenhouse. Plants that need "high" humidity are positioned closest to the fog-making machine. Finally, some orchids need special care every day or seasonally, and Bob likes to flag these plants' "special care" as a reminder.
2. Bob wants to track which species die before he can get them to market. If Bob doesn't have good luck holding certain kinds of orchids for quick resale, he doesn't want to buy them again!
3. Bob wants to track which species arrive from the wholesalers without buds or blooms. An out-of-bloom orchid doesn't sell for as much money as one that's in bloom, and Bob doesn't want to continue buying orchids from wholesalers who send him out-of-bloom plants.

4. Periodically, Bob needs to find out which kinds of orchids are his best sellers. Bob's customers run the range from first-time orchid buyers to expert growers, and their interests are different. He wonders whether some orchid groups sell much better than others.
5. Periodically, Bob needs to run a query to find out which orchids are the most profitable. This would be easy for orchids that Bob buys for quick resale—it's the difference between what he paid for certain plants and the amount for which he sold them.
6. Finally, Bob would like to determine his profit per supplier. He would like to see this in a final report.

Bob asks if you can identify other ways that a database might help his business operation. You say that you will make recommendations after examining his situation.

Case 3

Assignment 1 Creating the Database Design

In this assignment, you will design your database tables on paper, using a word-processing program. Pay close attention to the tables' logic and structure. Do not start your Access code (Assignment 2) before getting feedback from your instructor on Assignment 1. Keep in mind that you will need to look at what is required in Assignment 2 to design your fields and tables properly. You will also need to consider additional ways in which a database could help Bob. It's good programming practice to look at the outputs you need before designing your database. When designing the database, observe the following guidelines:

- First, determine the tables you'll need by listing on paper the name of each table and the fields that it should contain. Avoid data redundancy. Do not create a field if it could be created by a "calculated field" in a query.
- You'll need a transaction table. Avoid duplicating data.
- Think about some logical fields that you need to build into the tables.
- Document your tables by using the Table facility of your word processor. Your word-processed tables should resemble the format of the table in Figure 3-1.
- You must mark the appropriate key field(s). You can designate a key field by an asterisk (*) next to the field name. Keep in mind that some tables need a compound primary key to uniquely identify a record within a table.
- Print out the database design.

TABLE NAME	
Field Name	*Data Type (text, numeric, currency, etc.)*
...	...
...	...

Figure 3-1 Table design

Have your design approved before beginning Assignment 2; otherwise, you may need to redo Assignment 2.

ASSIGNMENT 2 CREATING A DATABASE AND DEVELOPING QUERIES AND A REPORT

In this assignment, you will first create database tables in Access and populate them with data. Then you will create five queries and one report. Name your database **ORCHIDS.mdb**.

Assignment 2A: Creating Tables in Access

In this part of the assignment, you will create your tables in Access. Observe the following guidelines:

- Type records into the tables, using the orchid information shown in Figure 3-2.
- Make up wholesaler information. Create at least 4 records. Because Bob buys at wholesale prices, he must buy at least 25 plants of any one kind from a wholesaler.
- Keep in mind that prices for each orchid may vary when being sold at a show. You can assume wholesale prices remain constant at $15. Use a retail selling price of between $25 and $45 per plant, but vary the price depending on the orchid group; for example, all Brassia orchids would resell for one price, and all Vanda orchids would sell for another, different price.
- For typing brevity, use sales from just one orchid show. If you type in 25 different records, you can copy and paste them repeatedly to make up at least 200 orchids being sold. Make sure that you vary some of the data. For example, make sure to note which plants were dead on arrival from the wholesaler; the majority of plants will be in bloom, but you will need to identify some that are not; choose some plants that did not sell at the show.
- Appropriately limit the size of the text fields; for example, a Supplier Number text field does not have to be the default 50 characters in length.
- Print all tables.

Orchid Name (group)	Light	Humidity	Heat	Special Care
Brassia	Intermediate	Intermediate	Intermediate/warm	Do not repot often
Cattleya	High	Intermediate	Intermediate/warm	Dry out between watering
Cymbidium	High	Intermediate	Cool	Need 6 weeks of cool temps in fall to bud
Dedrobium nobile	High	Intermediate/high	Warm	No water, no fertilizer, cool night temp for 1 mo
Dendrobium	High	Intermediate/high	Warm	
Miltonia	Intermediate	Intermediate	Intermediate	Keep moist, do not subject to >80 degrees
Oncidium	High/Intermediate	Intermediate	Intermediate/warm	No night temps <64 degrees
Paphiopedilum	Low	Intermediate/high	Intermediate/cool	Repot often, do not let dry out
Phalaenopsis	Low/intermediate	Intermediate/high	Warm	Do not let soil dry out or become too wet
Phragmipedium	Intermediate	Intermediate/high	Intermediate	Keep soil always moist, water daily
Vanda	High	High	Warm	Keep same conditions every day

Figure 3-2 Orchid information

Assignment 2B: Creating Queries and a Report

There are five queries and one report to generate, as outlined in the background of this case. Begin with the first query.

Query 1: Greenhouse Requirements

Since Bob has three greenhouses set up, he would like to figure out which orchids go into which house. Create a Parameter query that will prompt Bob to input the heat, light, and humidity of the orchid's requirements. Bob will use this information to sort the plants based on heat, then position them for light, and finally for humidity. For example, those requiring the most humid conditions could be placed near the fogger plant. Once those three parameters are entered, the query should display the Orchid Name and any extra Special Care notes, as shown in Figure 3-3, where the heat, light, and humidity entered were all intermediate.

Greenhouse Requirements : Select Query

Orchid Name (group)	Special Care
Miltonia	Keep moist, do not subject to >80 degrees

Figure 3-3 Greenhouse Requirements query, example output

Query 2: Orchids That Died

Bob has received some orchids from suppliers that were dead upon arrival. You noted those plants in the database. Create a query that lists those dead orchids by Item #, Orchid Name, and Supplier Name, as shown in Figure 3-4. Sort by Orchid Name and Supplier Name. Your data will differ (only a portion of the output is shown here).

Orchids That Died : Select Query

Item #	Orchid Name	Supplier Name
202	Brassia	Daniela's Blooms
70	Brassia	Daniela's Blooms
114	Brassia	Daniela's Blooms
48	Brassia	Daniela's Blooms
136	Brassia	Daniela's Blooms
26	Brassia	Daniela's Blooms
158	Brassia	Daniela's Blooms
4	Brassia	Daniela's Blooms
180	Brassia	Daniela's Blooms
92	Brassia	Daniela's Blooms
101	Phalaenopsis	Gaylord's Florals

Figure 3-4 Orchids That Died query, example output

Query 3: Orchids Not in Bloom

Similarly, Bob has received the occasional orchid or shipment of orchids that are not in bloom. Create a query similar to the previous query that shows those orchids not in bloom. Again, your data will differ, but the query should resemble that in Figure 3-5.

Orchids Not in Bloom : Select Query

Item #	Orchid Name	Supplier Name
101	Phalaenopsis	Gaylord's Florals
123	Phalaenopsis	Gaylord's Florals
145	Phalaenopsis	Gaylord's Florals
167	Phalaenopsis	Gaylord's Florals
189	Phalaenopsis	Gaylord's Florals
211	Phalaenopsis	Gaylord's Florals
13	Phalaenopsis	Gaylord's Florals
35	Phalaenopsis	Gaylord's Florals
57	Phalaenopsis	Gaylord's Florals
79	Phalaenopsis	Gaylord's Florals

Figure 3-5 Orchids Not in Bloom query, example output

Query 4: Best Sellers

Bob wants to identify his best-selling orchid groups. Create a query that counts the number of orchids he sells for this one show. (In an actual situation, you might choose a longer time period.) Display the orchid name and the number sold, listing them from highest seller to lowest seller. Make sure your column headings look good, as displayed in Figure 3-6.

Best Sellers : Select Query

Orchid Name	Number Sold
Brassia	70
Dendrobium nobile	30
Vanda	20
Paphiopedilum	20
Cattleya	20
Phragmipedium	10
Oncidium	10
Miltonia	10
Cymbium	10

Figure 3-6 Best Sellers query, example output

Query 5: Most Profitable Orchid Groups

Bob wants to make money, and hence he wants to track the profit of all his orchids sold. Create a calculated field that figures the total profit for all groups of orchids, from Brassia through Vanda. List in the output the number sold, orchid name, and the total profit, as shown in Figure 3-7. Your data will differ. Again, you will use the time period of this weekend's show.

Most Profitable Orchid Groups : Select Query

Number Sold	Orchid Name	Total Profit
70	Brassia	$1,050.00
40	Cattleya	$400.00
20	Vanda	$340.00
10	Phragmipedium	$230.00
10	Oncidium	$150.00
10	Miltonia	$150.00
20	Paphiopedilum	$120.00

Figure 3-7 Most Profitable Orchid Groups query, example output

Report: Profit by Supplier

Bob would like a report showing him how much profit he makes from each individual supplier. First, create a query to calculate the profit (retail sales price - wholesale cost), and then bring that into the Report Wizard and group on Supplier Name. Click the Summary Options button, and total the profit. Give the report a title, Profit by Supplier. Delete any bogus summary lines in the Design View, and make sure all fields are formatted for currency, if necessary. Although your data will differ, your output should resemble that in Figure 3-8. Again, use a one-weekend-show time period.

Profit by Supplier

Supplier Name	Orchid Name	Profit
Daniela's Blooms		
	Brassia	$1,050.00
Sum		$1,050.00
Dan's Wholesalers		
	Cattleya	$200.00
	Vanda	$340.00
Sum		$540.00

Figure 3-8 Profit by Supplier report, example output

Making a Presentation

Create a presentation for Bob. Explain the database and make suggestions for future work. Include the following:

- Describe the design of your database tables.
- Tell how to use the database, enter information, and run the queries and report.
- Bob asked you to think of additional ways that the database might help him to manage his business. Provide *one* additional data field, query, or report that will help Bob. For example, Bob might find it helpful to know which wholesalers' shipments were late—if he doesn't get plants in time, he can't sell them while they're in bloom! Bob might also like to know what his best sellers are by vendor.
- Explain how Bob might consider expanding the database if he ever decides to grow his own orchids for resale. You might also suggest how the database could be used if Bob were to expand his business operation in some other way that you think might be profitable.

Deliverables

1. Word-processed design of tables
2. Tables created in Access
3. Query 1: Greenhouse Requirements
4. Query 2: Orchids That Died
5. Query 3: Orchids Not in Bloom
6. Query 4: Best Sellers
7. Query 5: Most Profitable Orchid Groups
8. Report: Profit by Supplier
9. Presentation materials
10. Any other required tutorial printouts or tutorial disk or CD

Staple all pages together. Put your name and class number at the top of the page. Make sure your disk or CD is labeled.

The Dream Machine Rental Company

4

CASE

Designing a Relational Database to Create Tables, Queries, a Switchboard, and a Report

Preview

In this case, you'll design a relational database for a company that rents luxury automobiles, adding your own data fields, queries, or reports in addition to those required by your client. After your database design is completed and correct, you will create database tables and populate them with data.

Preparation

- Before attempting this case, you should have some experience in database design and in using Microsoft Access.
- Complete any part of Database Design Tutorial A that your instructor assigns.
- Complete any part of Access Tutorial B that your instructor assigns, or refer to the tutorial as necessary.
- Refer to Tutorial E as necessary.
- In this case, you will be using the following features of Access: Parameter queries, Summation queries, Select queries, Delete queries, Switchboards, and Reports.

BACKGROUND

In your university town, there is a company called The Dream Machine that rents luxury automobiles and provides limousine service. Beginning in 2000 with only two limousines, the owner has expanded his market to rent luxury self-drive cars. Because business has become very brisk at The Dream Machine, the company's owner has hired you to design and create a database system for the company.

The Dream Machine currently rents a varied fleet of vehicles:

- Two limousines, one white and one black, to cater to the customers' color preferences. Limousine rental includes the services of a uniformed chauffeur. Each limousine accommodates 10 to 12 persons.
- Two self-drive luxury cars: a new Bentley Continental GT Coupe that seats 5 persons, and a Rolls Royce Phantom that seats 5 to 6.
- Three self-drive high-performance sports cars: a red Ferrari that seats two persons, a Jaguar XJ that seats 4, and Range Rover that seats 5.
- Two other vehicles will be added to the fleet at the end of the month.

Here's how the business operates: When booking a rental, the company owner immediately checks on the customer's identification to ensure the customer is over 25 years of age (the company's insurance regulations for self-drive vehicles only). Customers are told that limousines are rented by the day, but a "day" is considered to have a limit of 6 hours. The self-drive vehicles are rented for an entire day, but they must be returned by 11 a.m. the following day or incur charges for another full day. (The afternoon is spent cleaning a vehicle for the next customer and restocking a limousine with niceties and requested items.) Because business is very busy, the owner requires that a deposit on the reservation be paid within 7 days of booking. High school students who book limousines for proms often make many installment payments, as do some brides. By contrast, those who rent high-performance sports cars are often repeat customers who pay the full amount in advance.

The owner is somewhat unsure of his exact needs, but he is sure that the database should perform the following tasks:

1. The owner would like to run a query to see which cars have been booked and the dates on which they are booked.
2. Sometimes, people call in to cancel their booking before making a deposit. The owner would like an easy way to delete a booking in a query.
3. Potential customers call in and ask for automobiles that can accommodate a certain number of people. Here, you will need to create a Parameter query that responds with the appropriate vehicle for the number of passengers required.
4. After reserving a vehicle, customers have 7 days in which to make a deposit on their reservation. They can continue paying for the rental in installments, or pay the entire amount. The owner would like a simple way to figure out how much money is remaining on every customer's reservation. You need to create a query that calculates the remaining balance.
5. The chauffeurs need to know when they have been hired to drive the limousines. The chauffeurs don't know how to use computers. The owner would like you to set up a Switchboard so the drivers can simply click on the button, type in their name, and see when they are needed for work, which is executed through a report.

Thus, your client has identified some—but not all—of his business needs. A good consultant provides extra value to clients by anticipating needs that the client has not yet considered. Consider your client's business operation. What other kinds of information could and should the database provide to your client?

ASSIGNMENT 1 CREATING THE DATABASE DESIGN

In this assignment, you will design your database tables on paper, using a word-processing program. Pay close attention to the tables' logic and structure. Do not start your Access code (Assignment 2) before getting feedback from your instructor on Assignment 1. Keep in mind that you will need to look at what is required in Assignment 2 to design your fields and tables properly. It's good programming practice to look at the required outputs before designing your database. When designing the database, observe the following guidelines:

- First, determine the kinds of information and data that the database should provide that the owner has NOT considered. Identify these data items.
- Second, determine the tables you'll need by listing on paper the name of each table and the fields that it should contain. Avoid data redundancy. Do not create a field if it could be created by a "calculated field" in a query.
- You'll need a transaction table. Avoid duplicating data.
- Think about a logical field that you need to build in to the tables.
- Document your tables by using the Table facility of your word processor. Your word-processed tables should resemble the format of the table in Figure 4-1.
- You must mark the appropriate key field(s). You can designate a key field by an asterisk (*) next to the field name. Keep in mind that some tables need a compound primary key to uniquely identify a record within a table.
- Print out the database design.

Case 4

TABLE NAME	
Field Name	*Data Type (text, numeric, currency, etc.)*
...	...
...	...

Figure 4-1 Table design

Have your design approved before beginning Assignment 2; otherwise, you may need to redo Assignment 2.

ASSIGNMENT 2 CREATING A DATABASE AND DEVELOPING QUERIES, A SWITCHBOARD, AND A REPORT

In this assignment, you will first create database tables in Access and populate them with data. Save your database as **THE DREAM MACHINE.mdb**. Next, you will create four queries, one switchboard, and one report. To assist the owner, you will go beyond these requirements, creating whatever additional queries or reports that you feel are warranted.

Assignment 2A: Creating Tables in Access

In this part of the assignment, you will create your tables in Access. Observe the following guidelines:

- Type records into the tables, using the vehicles of White Limousine, Black Limousine, Rolls Royce, Bentley, Ferrari, Jaguar, and Range Rover. As mentioned in the Background, two additional vehicles will be joining the fleet at the end of the month. Choose any two other vehicles that you like, and add them to the records.
- One limousine is black and one is white; the Ferrari is red. For all other vehicles, choose your own colors. Note the seating capacity for the two new vehicles that you choose.
- Using your own past experience or Internet research, set some daily rental rates for all vehicles.
- Assume that there are three drivers, and they are referred to by their last names: Lewis, Orney, and Fredricks.
- Create 15 bookings for a one-month period, some deposits, and some customers.
- Appropriately limit the size of the text fields; for example, a Phone Number field does not have to be the default 50 characters in length.
- Print all tables.

Assignment 2B: Creating Queries, a Switchboard, and a Report

There are four queries, one switchboard, and one report to generate, as outlined in the Background of this case. Begin with the first query.

Query 1: Car Bookings

The owner needs to know the dates on which each vehicle is booked. Create a query that displays that information, showing Car ID, Car Type, and Date/Time Rented. Sort the output on Car ID, Car Type, and Date/Time Rented. Your data will vary, and you may want to add additional information, but your core output should resemble that in Figure 4-2 (only a portion shows). Run the query and print the output.

Car Bookings : Select Query

Car ID	Car Type	Date/Time Rented
Bent	Bentley	5/12/2006 10:00:00 AM
Ferr	Ferrari	5/1/2006 5:00:00 PM
Ferr	Ferrari	5/16/2006 6:00:00 PM
Jag	Jaguar XJ	5/31/2006 8:00:00 PM
LimoB	Limo	5/1/2006 4:00:00 PM
LimoB	Limo	5/30/2006 1:00:00 PM

Figure 4-2 Car Bookings query

Query 2: Delete Booking

During the busy prom season, limousines are often booked temporarily but then the reservation is cancelled before the deposit is required. The owner would like to be able to delete such a record. Create a delete query to do just that. Save it as Delete Booking. Test the query by inputting a booking record and then deleting that record. Check the table to make sure the record was deleted. Print out the table and note with your pen which record was deleted.

Query 3: Number of Passengers

Customers call and ask for vehicles that can accommodate a certain number of passengers. Create a Parameter query that displays those vehicles that have the requested capacity. Depending on your data, if you typed in the requirement of 12 people, you might get the output as shown in Figure 4-3. Show the Car Type, Color, Number of Passengers, and Price per Day. Choose a capacity, run the query, and print the output.

Number of Passengers : Select Query

Car Type	Color	Number of Passenge	Price per Day
Limo	White	12	$720.00

Figure 4-3 Number of Passengers query

Query 4: Payments Remaining

Customers are required to make a deposit on their reservation within seven days of booking. They also can make payments on the reservation up to the time of rental, when the entire balance is due. The owner would like to run a query that shows the customer's Name, Address, Telephone, and any Payments Remaining. To do this, you must first make a query that calculates all the payments made on a given booking (because some bookings have more than one payment). Then, use that query in another query to calculate the payments remaining. Your data will vary, and you may have additional columns of information, but the core output of the query should resemble that shown in Figure 4-4. Print the query output.

Payments Remaining : Select Query

Name	Address	Telephone	Payments Remaining
Kathy Lopez	56 Myra Close, Avondale, PA 14556	(610)345-0987	$50.00
Angela Bernolli	4576 Kirkwood Hwy, Wilmington, DE 19808	(302)889-0987	$500.00
Kevin Barron	1 Close Circle, Hockessin, DE 19803	(302)998-7865	$220.00
Sam O'Hara	56 Church Road, Newark, DE 19711	(302)876-7765	$500.00

Figure 4-4 Payments Remaining query

Report: Drivers Required

The chauffeurs need an easy way to check on their driving assignments. They are not very computer literate, so you need to create a Switchboard they can use to simply click on their assignments. Before you make the switchboard, you must first make a report, based on a Parameter query, that prompts for the driver's name. After the driver's name is inputted, the resulting report should show the Driver Name, the Car ID, and the Date/Time Rented, as illustrated in Figure 4-5. Print the query output after inputting a sample driver's name.

Drivers Required

Driver Name	*Car ID*	*Date/Time Rented*
Fredricks	LimoW	5/29/2006 11:00:00 AM
Fredricks	LimoW	5/18/2006 11:00:00 AM
Fredricks	LimoW	5/5/2006 3:00:00 PM

Figure 4-5 Drivers Required report

Switchboard: The Dream Machine

After the report is created, you must create a switchboard for the drivers to use to run their Drivers Required report. Assign the button a title of Driver Assignment, as shown in Figure 4-6. Print the Switchboard interface by clicking File—Print while the Switchboard is visible on your screen.

Figure 4-6 The Dream Machine Switchboard

MAKING A PRESENTATION

Create a presentation for the owner. Explain the database and your additions to his requirements. You might also wish to make suggestions for future expansion of the database. Include the following in your presentation:

- Describe the design of your database tables.
- Tell how to use the database, per the owner's original requests: entering information, running the queries, using the switchboard, and generating the report.
- Explain the additional work that you've done for your client. This might include creating additional fields, queries, switchboards, or reports. Some suggestions might include the following: an Income Report; a Payroll Report for the chauffeurs; YES/NO fields for Damage to Vehicle, Repeat Customer, Late Return; an Actual Time Returned field; a section for noting extra charges for limo requests (French champagne, caviar, certain snacks).

Deliverables

1. Word-processed design of tables
2. Tables created in Access
3. Query 1: Car Bookings
4. Table: Bookings (after Delete Query)
5. Query 3: Number of Passengers
6. Query 4: Payments Remaining
7. Report: Drivers Required
8. Switchboard: The Dream Machine Switchboard
9. Presentation materials
10. Any other required tutorial printouts, tutorial disk, or CD

Staple all pages together. Put your name and class number at the top of the page. Make sure your disk or CD is labeled.

CASE

The Lawn Mower Repair Business

DESIGNING A RELATIONAL DATABASE TO CREATE TABLES, FORMS, QUERIES, AND A REPORT

PREVIEW

In this case, you'll design a relational database for a lawn mower repair business. After your database design is completed and correct, you will create database tables and populate them with data. Then you'll create two queries, three forms, a report, and a switchboard.

PREPARATION

- Before attempting this case, you should have some experience in database design and in using Microsoft Access.
- Complete any part of Database Design Tutorial A that your instructor assigns.
- Complete any part of Access Tutorial B that your instructor assigns, or refer to it as necessary.
- Refer to Tutorial E as necessary.
- In this case, you will be using the following features of Access: Summation queries, Forms and Subforms, the Report Wizard, and Switchboard.

Background

Johnny, an immigrant, believes that the USA is The Land of Opportunity. He's worked long hours for many years to make it so, and now he owns his own business, Johnny's Lawn Mower Repair Service. Johnny overhauls and repairs lawn mowers. When customers bring their mowers to Johnny's shop for an overhaul, he changes the oil, replaces the spark plugs, and sharpens the mower blades. If a mower is malfunctioning, he also fixes the lawn mower. His business has grown rapidly, and he needs a system that is better than his paper-based system to keep track of customers, jobs, and cash flow. Johnny has hired you to create a database system to help him streamline his business.

To get you started, Johnny calls you into the shop and introduces you to one of his friends, Mack, who is also one of his best customers. Mack runs a huge lawn-care service and brings all his mowers to Johnny's for servicing.

Mack (whose English is better than Johnny's) explains to you what happens when he drops off his lawn mowers for servicing. First, Johnny finds Mack's customer information in the file cabinet. Customer information includes all the information for each mower that a customer brings in for servicing. For example, Mack brings in 3 Honda mowers, 1 Toro mower, and 2 Sears mowers. Johnny then puts a numbered tag on each mower, and he starts the mower. If the mower doesn't start, he makes a note—usually the problem is simply the spark plug. Johnny also takes any additional notes about the general condition of each mower. After the mowers are serviced, Johnny notifies Mack to pick up the mowers. The cost of a regular overhaul (wash, oil change, spark plug change, and blade sharpening), is $30 per mower. Further servicing or repair is priced by the job. Johnny doesn't want information about additional servicing and repairs on the database at this stage, but he may want you to expand the project after he understands how to use the basic database.

When Johnny does work for a customer, he logs his servicing in a book. He notes the customer's number and how many mowers he services. He also writes down the number of each mower in this book. If the customer or mower is new, he assigns a new customer number and lawn mower number.

After this interview, you agree to the project and propose creating not only the database tables, but also other objects that would be helpful to the business.

Johnny will need to run two queries. To help Johnny keep track of accounts receivable, you will create a query that calculates the amount of money owed by customers. Keep in mind that some people drop off more than one mower to be serviced. Also, if Johnny knows his customer well, he will let him take the serviced mower away without paying, trusting that the customer will mail a check. The second query that you will create is one that will calculate the amount of cash that Johnny has received from a customer.

To help with data entry, you propose creating three forms:

1. You first will create a customer form that contains a mower subform. This will be handy for logging in a customer and documenting all the lawn mowers he/she drops off.
2. Next, you propose to create a form that documents the servicing of the mower, with a mower subform. The servicing is unique to the customer, so again, if the customer drops off more than one mower, the subform is necessary to display all the mowers.
3. Finally, you suggest a third form that logs all the cash receipts. Johnny's daughter, Holly, comes in on Saturdays, the busiest day. She writes a cash receipt for each transaction, noting the servicing number, the date, and the amount received. This goes onto a two-part form that has a unique number preprinted on the form.

Johnny also wants you to create a report that shows the different lawn mower types that each customer owns so he can keep the supplies he needs in stock.

To complete the project, you decide to create a switchboard that will allow Johnny and any other users of the database to easily access the forms and the report.

From your conversation with Johnny and Mack, you also learn that Johnny would, eventually, like to add lawn mower repairs to the database. Johnny is such a good mechanic, he is also considering doing small-engine repairs for equipment other than lawn mowers—and perhaps even hire his cousin from the Old Country to help him in the shop.

ASSIGNMENT 1 CREATING THE DATABASE DESIGN

In this assignment, you will design your database tables on paper, using a word-processing program. Pay close attention to the tables' logic and structure. Do not start your Access code (Assignment 2) before getting feedback from your instructor on Assignment 1. Keep in mind that you will need to look at what is required in Assignment 2 to design your fields and tables properly. It's good programming practice to look at the required outputs before designing your database. When designing the database, observe the following guidelines:

- First, determine the tables you'll need by listing on paper the name of each table and the fields that it should contain. Also, consider how the database might be used in the future, after Johnny gets more comfortable with it, to generate additional queries and reports. What additional tables and data fields might be needed? Plan these now.
- Avoid data redundancy. Do not create a field if it could be created by a "calculated field" in a query.
- You'll need a transaction table. Avoid duplicating data.
- Think about some logical fields that you need to build into the tables.
- Document your tables by using the Table facility of your word processor. Your word-processed tables should resemble the format of the table in Figure 5-1.
- You must mark the appropriate key field(s). You can designate a key field by an asterisk (*) next to the field name. Keep in mind that some tables need a compound primary key to uniquely identify a record within a table.
- Print out the database design.

TABLE NAME	
Field Name	*Data Type (text, numeric, currency, etc.)*
...	...
...	...

Figure 5-1 Table design

NOTE Have your design approved before beginning Assignment 2; otherwise, you may need to redo Assignment 2.

ASSIGNMENT 2 CREATING A DATABASE AND DEVELOPING FORMS, QUERIES, AND A REPORT

In this assignment, you will first create database tables in Access and populate them with data. Name and save your database as **MOWER.mdb**. Then you will create two queries, three forms, one report, and one switchboard.

Assignment 2A: Creating Tables in Access

In this part of the assignment, you will create your tables in Access. Use your imagination to create the data, but observe the following guidelines:

- To minimize typing, assume there are 6 customers. Use your classmates' names and addresses for data.
- Create data for 16 lawn mowers. You can use popular brands such as Sears, Toro, and Honda. Those 16 lawn mowers belong to the 6 customers.
- Each customer brings in lawn mowers for servicing. To simplify the data, assume each lawn mower just gets a basic service at $30. (Recall that Johnny wants to keep this initial system simple and not include extra repairs.) You can assume each customer's lawn mowers get serviced on the same day as they are brought in.
- For typing brevity, assume only one month's worth of data.
- Appropriately limit the size of the text fields; for example, a Customer Number field does not have to be the default 50 characters in length.
- Print all tables.

Assignment 2B: Creating Queries, Forms, a Report, and Switchboard

There are two queries, three forms, one report, and one switchboard to generate, as outlined in the background of this case. Bear in mind that you will be making recommendations for expanding the database's use. Ask your instructor how you should handle these future expansions. Begin with the first query.

Query 1: Money Owed by Customer

It's important to know how much money Johnny's customers owe him. Create a query that calculates accounts receivable. First, you must create a query that counts the number of lawn mowers each customer leaves at the shop. You will save that query and use it in another query to calculate the charge for the lawn mower servicing. If you are unsure about how to use a query as a basis for another query, use the Help facility in Access. Keep in mind that each lawn mower costs $30 for a full service. List the Customer Number, Last Name, Charges for Servicing, Receipts of Cash Paid, and Remaining Money Owed. Save the query as Money Owed by Customer. Sort on Customer Number. Your data will differ, but the format of your output should resemble that in Figure 5-2. Print the output.

Customer Number	Last Name	Charges for Servicing	Receipts of Cash Paid	Remaining Money Owed
C101	Brady	$150.00	$30.00	$120.00
C102	Monk	$60.00	$30.00	$30.00
C103	Davidson	$180.00	$60.00	$120.00
C104	Wright	$30.00	$30.00	$0.00

Figure 5-2 Money Owed by Customer

Query 2: Cash Received

You would also like to see a query that shows all the money taken in thus far from each customer. Create a query that lists total money paid per customer, showing the customer number and cash received. Save the query as Cash Received. Your output should resemble the format of that in Figure 5-3. Print the output.

Cash Received : Select Query

Customer Number	Cash Received
C101	$30.00
C102	$30.00
C103	$60.00
C104	$30.00

Figure 5-3 Cash Received

Form 1: Customer

Next, create a form that would make it easy for Johnny to check in new customers. This form should consist of a main form, based on your CUSTOMER table, and a subform, based on the MOWER table. Keep in mind that some customers drop off more than one lawn mower. Your data may differ, but your form will resemble the one shown in Figure 5-4. Print one record.

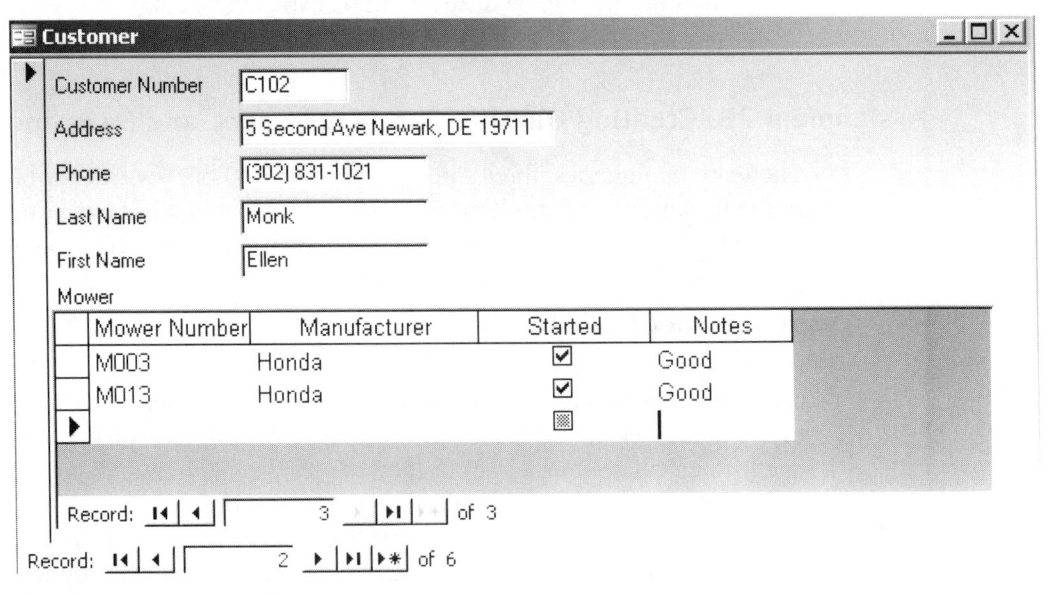

Figure 5-4 Customer form

Form 2: Servicing

The next form should help Johnny to fill out his service records. The form should be based on your SERVICE table with a mower subform, because some customers' service includes more than one mower. This will allow Johnny to type in the service number, the customer number, and then the date of service. Your data will differ, but your form will resemble the one shown in Figure 5-5. Print one record.

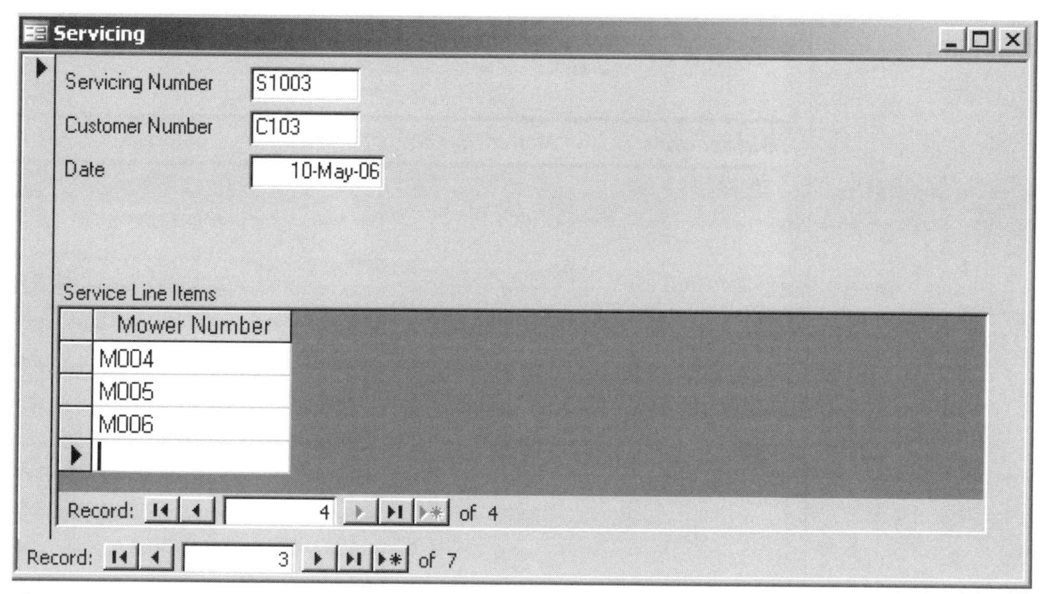

Figure 5-5 Servicing Mowers form

Form 3: Cash Receipts

It would be handy to be able to record the cash receipts in a form. Holly does this by hand, but would prefer to type it directly into the database via a form. Create a form for that purpose, as shown in Figure 5-6. Print one record.

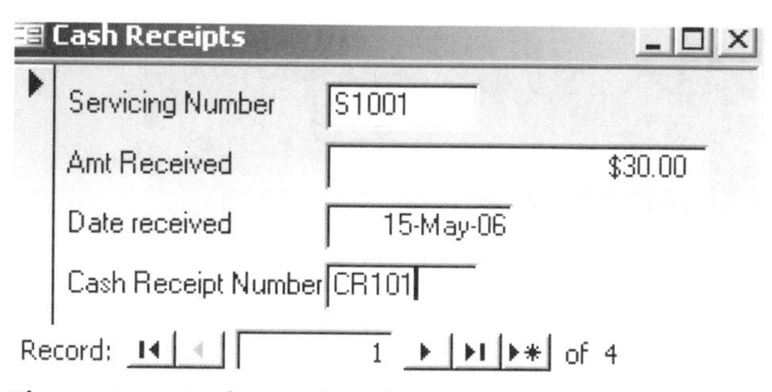

Figure 5-6 Cash Receipts form

Report: Mower Types

Johnny wants to buy adequate supplies for each type of lawn mower he services. Because he has many repeat customers—many of whom own lawn-care businesses—he would like to see a report of customers for each type of mower. Using first a query and then the Report Wizard, for each mower show the Manufacturer, Mower Number, customer's Last Name, customer's Address, and customer's Phone number. Group the report on Manufacturer, and title the report Mower Types. Your data will differ, but the output should look like that in Figure 5-7. Print the report.

Mower Types

Manufacturer	*Mower Number*	*LastName*	*Address*	*Phone*
Honda				
	M015	Davidson	9 Post Pl Newark, DE 19711	(302) 831-1324
	M013	Monk	5 Second Ave Newark, DE 19711	(302) 831-1021
	M010	Smith	8 South St Newark, DE 19711	(302) 831-1221
	M003	Monk	5 Second Ave Newark, DE 19711	(302) 831-1021
	M005	Davidson	9 Post Pl Newark, DE 19711	(302) 831-1324
Sears				
	M014	Davidson	9 Post Pl Newark, DE 19711	(302) 831-1324
	M012	Brady	8 Main St Newark, DE 19711	(302) 831-1001
	M002	Brady	8 Main St Newark, DE 19711	(302) 831-1001
	M004	Davidson	9 Post Pl Newark, DE 19711	(302) 831-1324

Figure 5-7 Mower Types report

Switchboard: Mower Servicing

To make the project look professional, create a switchboard that shows buttons for each form and the report. Your final switchboard should look something like that shown in Figure 5-8. Print the interface by clicking File—Print.

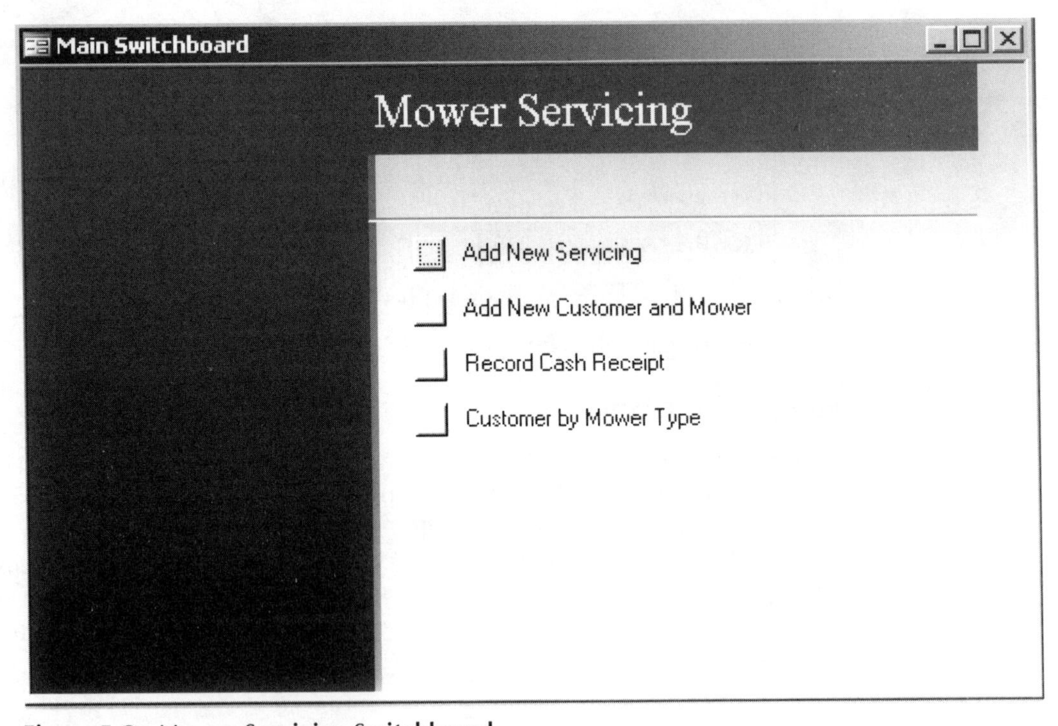

Figure 5-8 Mower Servicing Switchboard

Making a Presentation

Create a presentation for Johnny. Explain the database and make suggestions for future work. Include the following:

- Describe the design of your database tables.
- Tell how to use the database: enter information, run the queries, generate the report, and use the switchboard.
- Explain how Johnny might streamline the process to service more lawn mowers more efficiently using this database and further enhancements. Ask your instructor how detailed these changes should be. You might be asked to redesign the database and create further queries.
- Suggest ways to expand the database to include other types of servicing and repair of other equipment.

Deliverables

1. Word-processed design of tables
2. Tables created in Access
3. Query 1: Money Owed by Customers
4. Query 2: Cash Received
5. Form 1: Customer
6. Form 2: Servicing
7. Form 3: Cash Receipts
8. Report: Customer by Mower Type
9. Switchboard: Mower Servicing
10. Presentation materials
11. Any other required tutorial printouts or tutorial disk or CD

Staple all pages together. Put your name and class number at the top of the page. Make sure your disk or CD is labeled.

6

CASE

Challenge Case: The Online Dating Service

Designing a Relational Database to Create Tables, a Form, Queries, and a Report

Preview

In this case, you'll design a relational database for a company that runs an online dating service. After your database design is completed and correct, you will create database tables and populate them with data.

Unlike previous cases, you will not be shown the format of your output. You must determine the format of the output and then create your database tables accordingly.

Preparation

- Before attempting this case, you should have experience in database design and in using Microsoft Access.
- Complete any part of Database Design Tutorial A that your instructor assigns.
- Complete any part of Access Tutorial B that your instructor assigns, or refer to it as necessary.
- Refer to Tutorial E as necessary.
- In this case, you will be using the following features of Access: Forms, Select queries, Parameter queries, Update queries, Delete queries, and Switchboards

Background

Online dating is one of the Internet's growth areas. In 2004, it was estimated that 40 million people will participate in online dating. One of your college friends, Sarah "Single" St. John, has decided to get in on this growth market and start an online dating service. Because you are such an expert in Microsoft Access, she calls you to set up the business's database.

Sarah stipulates that the initial form that the client fills out should not be too lengthy. Although you can achieve better date matching with intricate forms, clients get bored with filling them out. So, the initial pass for a new client will be to fill out an abbreviated form. Later, they can opt to fill out a more detailed form to get more accurate matching. On the form, Sarah would like to see the following items:

1. Name, address, telephone number
2. Gender
3. Age, in terms of date of birth, to eliminate updating old records.
4. Religion
5. Occupation
6. Height
7. Leisure activities, limited to three major ones, ranked in order of preference.
8. Income, in a broad range bracket.
9. Smoker?
10. Hair color
11. Children?

You ask Sarah why she does not want clients to post a photograph of themselves. Sarah claims that photos on the Internet can be very deceiving. Photo content can also vary—some photos show a person's whole body, and some show just the head. Photos can also be very outdated. In addition, some of her clients do not have digital cameras, and that might cause clients to decide not to join.

Sarah would like you to set up a form for inputting data, a number of queries, and a report for the service. There should be a series of queries that filter the tables to find people with similar interests and traits. For example, it would be efficient to have one query that matches or sorts people based on their religious preference, another that matches or sorts people on their top three leisure activities, and so on. Also, one of Sarah's good friends is a nurse and is always concerned about the number of people who smoke. Sarah agrees to give her friend some basic statistics on the number of smokers and non-smokers in her dating database.

Sometimes, a client's information (phone number, etc.) changes. There needs to be an easy way to update information on a client. Also, some clients will be dropped from the database if they have found a mate or if they are not truthful about their information. An Update query would be useful for changing information. A Delete query could be used to delete clients who have found a mate or have lied on their application.

Finally, create a switchboard to give the system a neat interface for inputting new member data.

ASSIGNMENT 1 CREATING THE DATABASE DESIGN

In this assignment, you will design your database tables on paper, using a word-processing program. Pay close attention to the tables' logic and structure. Do not start your Access code (Assignment 2) before getting feedback from your instructor on Assignment 1. Keep in mind that you will need to determine what is required in Assignment 2 to design your fields and tables properly. It's good programming practice to look at the required outputs before designing your database. When designing the database, observe the following guidelines:

- First, determine the output of Query 7. Then determine the tables you'll need by listing on paper the name of each table and the fields that it should contain. Avoid data redundancy. Do not create a field if it could be created by a "calculated field" in a query.
- Think about some logical fields that you might need to build into the tables.
- Document your tables by using the Table facility of your word processor. Your word-processed tables should resemble the format of the table in Figure 6-1.
- You must mark the appropriate key field(s). You can designate a key field by an asterisk (*) next to the field name. Keep in mind that some tables need a compound primary key to uniquely identify a record within a table.
- Print out the database design.

TABLE NAME	
Field Name	***Data Type (text, numeric, currency, etc.)***
...	...
...	...

Figure 6-1 Table design

Have your design approved before beginning Assignment 2; otherwise, you may need to redo Assignment 2.

ASSIGNMENT 2 CREATING A DATABASE AND DEVELOPING A FORM, QUERIES, AND A REPORT

In this assignment, you will first create database tables in Access and populate them with data. Name and save your database as **Dating Service.mdb.** Then you will create one form, seven queries, and one report.

Assignment 2A: Creating Tables in Access

In this part of the assignment, you will create your tables in Access. Observe the following guidelines:

- Type records into the tables, asking your classmates for help. Ask five of your classmates to create a profile of two perfect dates, for a total of 10 profiles. Make sure that they choose all the parameters that Sarah is tracking: gender, date of birth, religion, occupation, height, income, smoking status, hair color, whether or not they have children, and their three favorite leisure activities—ranked in order of importance. Make up names, addresses, and telephone numbers for these 10 records.

- Assign 3 leisure activities to each, limiting the choice to 8 activities.
- Assign a religion to each, using at least 4 religions.
- Use popular hair colors, and assign a varied range of ages.
- Allow some of the customers to have found mates through your service.
- Appropriately limit the size of the text fields; for example, a Phone Number field does not have to be the default 50 characters in length.
- Print all tables.

Assignment 2B: Creating a Form, Queries, and a Report

You will be creating one form, seven queries, and one report, as outlined in the background of this case. Begin with the form.

Form: Customer Data

Sarah would like to be able to type a new customer's data and traits directly into the database. Create a form that would allow her to do that easily.

Query 1: One Trait

Create a query that filters the data based on one trait, such as a person's favorite leisure activity. Suppose that someone calls up the dating service and is very interested in meeting someone who really likes to kayak. Set up a query to find those people. Sort by gender.

Query 2: Two Traits

Now make a query that filters for two traits. This time, assume that a customer calls up and wants to find a mate who likes to read and is over 50 years old. Sort by gender.

Query 3: Three Traits

Repeat Query 2, but now filter for three traits. Assume that a customer calls in and wants to find a mate under age 35, with children, who likes to bicycle. Sort by gender, then by age.

Query 4: Number of Smokers

Make a query that counts the number of smokers in the database and counts the number of non-smokers. Make sure your column headings look good.

Query 5: Update Query

Customers' information changes. Create an Update query that allows a specific customer to change his/her phone number. You can make that update query prompt for the new telephone number. Print out the updated table and circle the changed records.

Query 6: Delete Query

Customers rarely decide to drop out of the dating service. More likely, they find a mate through the service and then no longer want to be called to date new people. Create a Delete query, based on someone's record, that deletes him or her from the appropriate table. Again, you can make this query simply prompt for the customer's name or number. Print the table and write in the record you deleted at the top.

Query 7: (Your Choice)

Think of another query that Sarah might find useful. Create that query, run it, and print the results. On the results printout, describe to your instructor how you set up the query and what question it answers for Sarah.

Switchboard: The Dating Service

Create a switchboard to give the system a professional look. Include the form to input new member data in your switchboard, and call the switchboard The Dating Service. Print the interface to the switchboard.

Making a Presentation

Create a presentation to the owner. Explain the database and make suggestions for future work. Include the following:

- Describe the design of your database tables.
- Tell how to use the database, enter information via the form, and how to run the queries and the report.
- Describe Query 7 in detail.

Deliverables

1. Word-processed design of tables
2. Tables created in Access
3. One sample record printed from the form
4. Query 1: One Trait
5. Query 2: Two Traits
6. Query 3: Three Traits
7. Query 4: Number of Smokers
8. Query 5: (Update Query) Updated Table
9. Query 6: (Delete Query) Table with Deleted With Deleted Record Noted
10. Query 7: (Your Choice)
11. Switchboard interface
12. Presentation materials
13. Any other required tutorial printouts or tutorial disk or CD

Staple all pages together. Put your name and class number at the top of the page. Make sure your disk or CD is labeled.

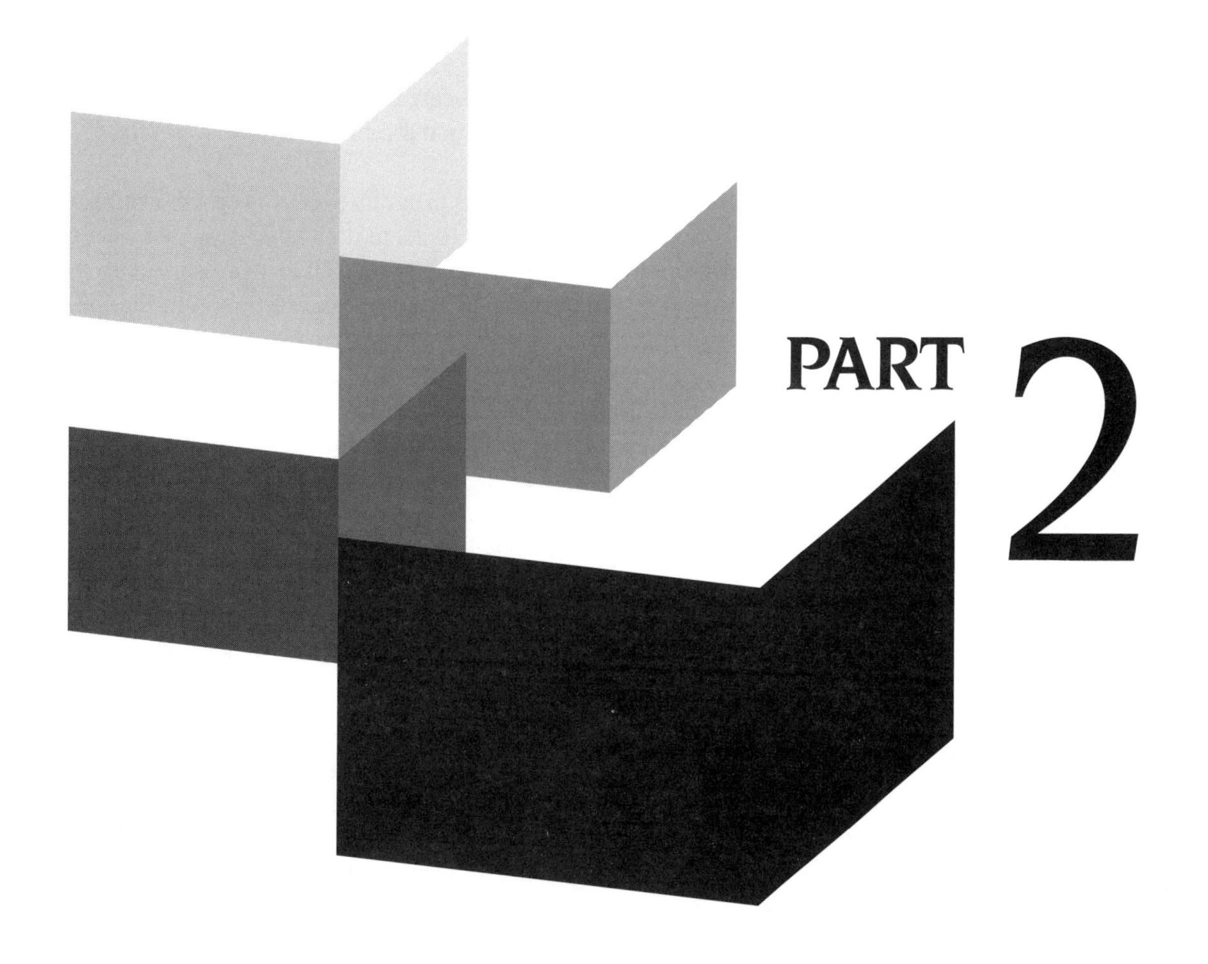

Decision Support Cases Using Excel Scenario Manager

Building a Decision Support System in Excel

A **decision support system (DSS)** is a computer program that can represent, either mathematically or symbolically, a problem that a user needs to solve. Such a representation is, in effect, a model of a problem.

Here's how a DSS program works: The DSS program accepts input from the user or looks at data in files on disk. Then, the DSS program runs the input and any other necessary data through the model. The program's output is the information the user needs to solve a problem. Some DSS programs even recommend a solution to a problem.

A DSS can be written in any programming language that lets a programmer represent a problem. For example, a DSS could be built in a third-generation language, such as Visual Basic, or in a database package, such as Access. A DSS could also be written in a spreadsheet package, such as Excel.

The Excel spreadsheet package has standard built-in arithmetic functions as well as many statistical and financial functions. Thus, many kinds of problems—such as those in accounting, operations, or finance—can be modeled in Excel.

This tutorial has the following four sections:

1. **Spreadsheet and DSS Basics:** In this section, you'll learn how to create a DSS program in Excel. Your program will be in the form of a cash flow model. This will give you practice in spreadsheet design and in building a DSS program.
2. **Scenario Manager:** In this section, you'll learn how to use an Excel tool called the Scenario Manager. With any DSS package, one problem with playing "what if" is this: Where do you physically record the results from running each set of data? Typically, a user just writes the inputs and related results on a piece of paper. Then—ridiculously enough—the user might have to input the data *back* into a spreadsheet for further analysis! The Scenario Manager solves that problem. It can be set up to capture inputs and results as "scenarios," which are then nicely summarized on a separate sheet in the Excel workbook.
3. **Practice Using Scenario Manager:** You are invited to work on a different problem, a case using the Scenario Manager.
4. **Review of Excel Basics:** This section reviews the information you'll need to do the spreadsheet cases that follow this tutorial.

SPREADSHEET AND DSS BASICS

Assume it is late in 2005, and that you are trying to build a model of what a company's net income (profit) and cash flow will be in the next two years (2006 and 2007). This is the problem: to forecast net income and cash flow in those years. Assume that knowing these forecasts would help to answer some question or make some decision. After researching the problem, you decide that the estimates should be based on three things: (1) 2005 results, (2) estimates of the underlying economy, and (3) the cost of products the company sells.

The model will use an income statement and cash flow framework. The user can input values for two possible states of the economy in years 2006-2007: an "O" for an Optimistic outlook or a "P" for a Pessimistic outlook. The state of the economy is expected to affect the number of units the company can sell as well as the unit selling price: In a good "O" economy, more units can be sold at a higher price. The user can also input values for two possible cost-of-goods-sold price directions: a "U" for Up or a "D" for Down. A "U" means that the cost of an item sold will be higher than in 2005; a "D" means that it will be less.

Presumably, the company will do better in a good economy and with lower input costs—but how much better? The relationships are too complex to assess in one's head, but the software model can easily assess the relationships. Thus, the user can play "what if" with the input variables and note the effect on net income and year-end cash levels. For example, a user can ask, "What if the economy is good and costs go up? What will net income and cash flow be in that case? What would happen if the economy is down and costs go down? What would be the company's net income and cash flow in that case?" With an Excel software model available, the answers are easily quantified.

Organization of the DSS Model

Your spreadsheets should have the following sections, which will be noted in boldface type throughout this tutorial and in the Excel cases that follow it:

- **CONSTANTS**
- **INPUTS**
- **SUMMARY OF KEY RESULTS**
- **CALCULATIONS** (of values that will be used in the INCOME STATEMENT AND CASH FLOW STATEMENT)
- **INCOME STATEMENT AND CASH FLOW STATEMENT**

Here, as an extended illustration, a DSS model is built for the forecasting problem previously described. Let's look at each spreadsheet section. Figures C-1 and C-2 show how to set up the spreadsheet.

	A	B	C	D
1	**TUTORIAL EXERCISE**			
2				
3	**CONSTANTS**	**2005**	**2006**	**2007**
4	TAX RATE	NA	0.33	0.35
5	NUMBER OF BUSINESS DAYS	NA	300	300
6				
7	**INPUTS**	**2005**	**2006**	**2007**
8	ECONOMIC OUTLOOK (O = OPTIMISTIC; P=PESSIMISTIC)	NA		NA
9	PURCHASE-PRICE OUTLOOK (U = UP; D = DOWN)	NA		NA
10				
11	**SUMMARY OF KEY RESULTS**	**2005**	**2006**	**2007**
12	NET INCOME AFTER TAXES	NA		
13	END-OF-THE-YEAR CASH ON HAND	NA		
14				
15	**CALCULATIONS**	**2005**	**2006**	**2007**
16	NUMBER OF UNITS SOLD IN A DAY	1000		
17	SELLING PRICE PER UNIT	7.00		
18	COST OF GOODS SOLD PER UNIT	3.00		
19	NUMBER OF UNITS SOLD IN A YEAR	NA		

Figure C-1 Tutorial skeleton 1

	A	B	C	D
21	**INCOME STATEMENT AND CASH FLOW STATEMENT**	**2005**	**2006**	**2007**
22	BEGINNING-OF-THE-YEAR CASH ON HAND	NA		
23				
24	SALES (REVENUE)	NA		
25	COST OF GOODS SOLD	NA		
26	INCOME BEFORE TAXES	NA		
27	INCOME TAX EXPENSE	NA		
28	NET INCOME AFTER TAXES	NA		
29				
30	END-OF-THE-YEAR CASH ON HAND (BEGINNING-OF-THE-YEAR CASH, PLUS NET INCOME AFTER TAXES)	10000		

Figure C-2 Tutorial skeleton 2

The CONSTANTS Section

This section records values that are used in spreadsheet calculations. In a sense, the constants are inputs, except that they do not change. In this tutorial, constants are TAX RATE and the NUMBER OF BUSINESS DAYS.

The INPUTS Section

The inputs are for the ECONOMIC OUTLOOK and PURCHASE-PRICE OUTLOOK (manufacturing input costs). Inputs could conceivably be entered for *each year* covered by the model (here, 2006 and 2007). This would let you enter an "O" for 2006's economy in one cell and a "P" for 2007's economy in another cell. Alternatively, one input for the two-year period could be entered in one cell. For simplicity, this tutorial uses the *latter* approach.

The SUMMARY OF KEY RESULTS Section

This section will capture 2006 and 2007 NET INCOME AFTER TAXES (profit) for the year and END-OF-THE-YEAR CASH ON HAND, which are (assume) the two relevant outputs of this model. The summary merely repeats, in one easy-to-see place, results that are shown in otherwise widely spaced places in the spreadsheet. This just makes the answers easier to see all at once. (It also makes it easier to graph results later.)

The CALCULATIONS Section

This area is used to compute the following data:

1. The NUMBER OF UNITS SOLD IN A DAY (a function of a certain 2005 value and of the input economic outlook)
2. The SELLING PRICE PER UNIT (similarly derived)
3. The COST OF GOODS SOLD PER UNIT (a function of a 2005 value and of the purchase-price outlook)
4. The NUMBER OF UNITS SOLD IN A YEAR (the number of units sold in a day times the number of business days)

These formulas could be embedded in the **INCOME STATEMENT AND CASH FLOW STATEMENT** section. Doing that would, however, cause the expressions there to be complex and difficult to understand. Putting the intermediate calculations into a separate **CALCULATIONS** section breaks up the work into modules. This is good form because it simplifies your programming.

The INCOME STATEMENT AND CASH FLOW STATEMENT Section

This is the "body" of the spreadsheet. It shows the following:

1. BEGINNING-OF-THE-YEAR CASH ON HAND, which equals cash at the end of the *prior* year.
2. SALES (REVENUE), which equals the units sold in the year times the unit selling price.
3. COST OF GOODS SOLD, which is units sold in the year times the price paid to acquire or make the unit sold.
4. INCOME BEFORE TAXES, which equals sales less total costs.
5. INCOME TAX EXPENSE, which is zero if there are losses; otherwise, it is the pre-tax margin times the constant tax rate. (This is sometimes called INCOME TAXES.)
6. NET INCOME AFTER TAXES, which equals income before taxes less income tax expense.
7. END-OF-THE-YEAR CASH ON HAND is beginning-of-the-year cash on hand plus net income. (In the real world, cash flow estimates must account for changes in receivables and payables. In this case, assume that sales are collected immediately—i.e., there are no receivables or bad debts. Assume also that suppliers are paid immediately—i.e., that there are no payables.)

Construction of the Spreadsheet Model

Next, let's work through the following three steps to build your spreadsheet model:

1. Make a "skeleton" of the spreadsheet, and call it **TUTC.xls**.
2. Fill in the "easy" cell formulas.
3. Enter the "hard" spreadsheet formulas.

Make a Skeleton

Your first step is to set up a skeleton worksheet. This should have headings, text string labels, and constants—but no formulas.

To set up the skeleton, you must get a grip on the problem, *conceptually*. The best way to do that is to work *backward* from what the "body" of the spreadsheet will look like. Here, the body is the **INCOME STATEMENT AND CASH FLOW STATEMENT** section. Set that up, in your mind or on paper, then do the following:

- Decide what amounts should be in the **CALCULATIONS** section. In this tutorial's model, SALES (revenue) will be NUMBER OF UNITS SOLD IN A DAY times SELLING PRICE PER UNIT, in the income statement. You will calculate the intermediate amounts (NUMBER OF UNITS SOLD IN A YEAR and SELLING PRICE PER UNIT) in the **CALCULATIONS** section.
- Set up the **SUMMARY OF KEY RESULTS** section by deciding what *outputs* are needed to solve the problem. The **INPUTS** section should be reserved for amounts that can change—the controlling variables—which are the ECONOMIC OUTLOOK and the PURCHASE-PRICE OUTLOOK.
- Use the **CONSTANTS** section for values that you will need to use, but that are not in doubt, i.e., you will not have to input them or calculate them. Here, the TAX RATE is a good example of such a value.

AT THE KEYBOARD

Type in the Excel skeleton shown in Figures C-1 and C-2.

A designation of "NA" means that a cell will not be used in any formula in the worksheet. The 2005 values are needed only for certain calculations, so for the most part, the 2005 column's cells just show "NA." (Recall that the forecast is for 2006 and 2007.) Also be aware that you can "break" a text string in a cell by pressing the keys Alt and Enter at the same time at the break point. This makes the cell "taller." Formatting of cells to show centered data and creation of borders is discussed at the end of this tutorial.

Fill in the "Easy" Formulas

The next step is to fill in the "easy" formulas. The cells affected (and what you should enter) are discussed next.

To prepare, you should format the cells in the **SUMMARY OF KEY RESULTS** section for no decimals. (Formatting for numerical precision is discussed at the end of this tutorial.) The **SUMMARY OF KEY RESULTS** section just "echoes" results shown in other places. The idea is that C28 holds the NET INCOME AFTER TAXES. You want to echo that amount in C12. So, the formula in C12 is =C28. Translation: "Copy what is in C28 into C12." It's that simple.

With the insertion point in C12, the contents of that cell—in this case the formula =C28—shows in the editing window, which is above the lettered column indicators, as shown in Figure C-3.

C12	fx =C28		
	A	B	C	D
11	SUMMARY OF KEY RESULTS	2005	2006	2007
12	NET INCOME AFTER TAXES	NA	0	
13	END-OF-THE-YEAR CASH ON HAND	NA		

Figure C-3 Echo 2006 NET INCOME AFTER TAXES

At this point, C28 is empty (and thus has a zero value), but that does not prevent you from copying. So, copy cell C12's formula to the right, to cell D12. (The copy operation does not actually "copy.") Copying puts =D28 into D12, which is what you want. (Year 2007's NET INCOME AFTER TAXES is in D28.)

To perform the Copy operation, use the following steps:

1. Select (click in) the cell (or range of cells) that you want to copy.
2. Choose **Edit—Copy**.
3. Select the cell (or cell range) to be copied to by clicking (and then dragging if the range has more than one cell).
4. Press the **Enter** key.

END-OF-THE-YEAR CASH ON HAND for 2006 cash is in cell C30. Echo the cash results in cell C30 to cell C13. (Put =C30 in cell C13, as shown in Figure C-4.) Copy the formula from C13 to D13.

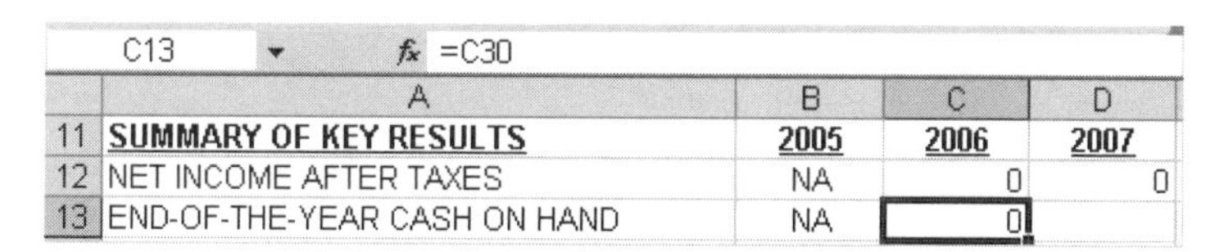

C13 fx =C30

	A	B	C	D
11	SUMMARY OF KEY RESULTS	2005	2006	2007
12	NET INCOME AFTER TAXES	NA	0	0
13	END-OF-THE-YEAR CASH ON HAND	NA	0	

Figure C-4 Echo 2006 END-OF-THE-YEAR CASH ON HAND

At this point, the **CALCULATIONS** section formulas will not be entered because they are not all "easy" formulas. Move on to the easier formulas in the **INCOME STATEMENT AND CASH FLOW STATEMENT** section, as if the calculations were already done. Again, the empty **CALCULATIONS** section cells in this section do not stop you from entering formulas. You should now format the cells in the **INCOME STATEMENT AND CASH FLOW STATEMENT** section for zero decimals.

BEGINNING-OF-THE-YEAR CASH ON HAND is the cash on hand at the end of the *prior* year. In C22 for the year 2006, type =B30. See the "skeleton" you just entered, as shown in Figure C-5. Cell B30 has the END-OF-THE-YEAR CASH ON HAND for 2005.

C22	=B30		
	A	B	C	D

	A	B	C	D
21	**INCOME STATEMENT AND CASH FLOW STATEMENT**	**2005**	**2006**	**2007**
22	BEGINNING-OF-THE-YEAR CASH ON HAND	NA	10000	
23				
24	SALES (REVENUE)	NA		
25	COST OF GOODS SOLD	NA		
26	INCOME BEFORE TAXES	NA		
27	INCOME TAX EXPENSE	NA		
28	NET INCOME AFTER TAXES	NA		
29				
30	END-OF-THE-YEAR CASH ON HAND (BEGINNING-OF-THE-YEAR CASH, PLUS NET INCOME AFTER TAXES)	10000		

Figure C-5 Echo of END-OF-THE-YEAR CASH ON HAND for 2005 to BEGINNING-OF-THE-YEAR CASH ON HAND for 2006

Copy the formula in cell C22 to the right. SALES (REVENUE) is just NUMBER OF UNITS SOLD IN A YEAR times SELLING PRICE PER UNIT. In cell C24, enter =C17*C19, as shown in Figure C-6.

C24 =C17*C19

	A	B	C	D
15	**CALCULATIONS**	**2005**	**2006**	**2007**
16	NUMBER OF UNITS SOLD IN A DAY	1000		
17	SELLING PRICE PER UNIT	7.00		
18	COST OF GOODS SOLD PER UNIT	3.00		
19	NUMBER OF UNITS SOLD IN A YEAR	NA		
20				
21	**INCOME STATEMENT AND CASH FLOW STATEMENT**	**2005**	**2006**	**2007**
22	BEGINNING-OF-THE-YEAR CASH ON HAND	NA	10000	
23				
24	SALES (REVENUE)	NA	0	
25	COST OF GOODS SOLD	NA		

Figure C-6 Enter the formula to compute 2006 SALES

The formula C17*C19 multiplies units sold for the year times the unit selling price. (Cells C17 and C19 are empty now, which is why SALES shows as zero after the formula is entered.) Copy the formula to the right, to D24.

COST OF GOODS SOLD is handled similarly. In C25, type =C18*C19. This equals NUMBER OF UNITS SOLD IN A YEAR times COST OF GOODS SOLD PER UNIT. Copy to the right.

In cell C26, the formula for INCOME BEFORE TAXES is =C24–C25. Enter the formula. Copy to the right.

In the United States, income taxes are only paid on positive income before taxes. In cell C27, shown in Figure C-7, the INCOME TAX EXPENSE is zero if the INCOME BEFORE TAXES is zero or less; else, INCOME TAX EXPENSE equals the TAX RATE times the INCOME BEFORE TAXES. The TAX RATE is a constant (in C4). An =IF() statement is needed to express this logic:

IF(INCOME BEFORE TAXES is <= 0, put zero tax in C27,
else in C27 put a number equal to multiplying the
TAX RATE times the INCOME BEFORE TAXES)

C26 stands for the concept INCOME BEFORE TAXES, and C4 stands for the concept TAX RATE. So, in Excel, substitute those cell addresses:

=IF(C26 <= 0, 0, C4 * C26)

Copy the income tax expense formula to the right.

In cell C28, NET INCOME AFTER TAXES is just INCOME BEFORE TAXES less INCOME TAX EXPENSE: =C26-C27. Enter and copy to the right.

The END-OF-THE-YEAR CASH ON HAND is BEGINNING-OF-THE-YEAR CASH ON HAND plus NET INCOME AFTER TAXES. In cell C30, enter =C22+C28. The **INCOME STATEMENT AND CASH FLOW STATEMENT** section at that point is shown in Figure C-7. Then, copy the formula to the right.

	A	B	C	D
21	**INCOME STATEMENT AND CASH FLOW STATEMENT**	**2005**	**2006**	**2007**
22	BEGINNING-OF-THE-YEAR CASH ON HAND	NA	10000	10000
23				
24	SALES (REVENUE)	NA	0	0
25	COST OF GOODS SOLD	NA	0	0
26	INCOME BEFORE TAXES	NA	0	0
27	INCOME TAX EXPENSE	NA	0	0
28	NET INCOME AFTER TAXES	NA	0	0
29				
30	END-OF-THE-YEAR CASH ON HAND (BEGINNING-OF-THE-YEAR CASH, PLUS NET INCOME AFTER TAXES)	10000	10000	

Figure C-7 Status of INCOME STATEMENT AND CASH FLOW STATEMENT

Put in the "Hard" Formulas

The next step is to finish the spreadsheet by filling in the "hard" formulas.

AT THE KEYBOARD

First, in C8 enter an "O" (no quotation marks) for OPTIMISTIC, and in C9 enter "U" (no quotation marks) for UP. There is nothing magic about these particular values—they just give the worksheet formulas some input to process. Recall that the inputs will cover both 2006 and 2007. Make sure "NA" is in D8 and D9, just to remind yourself that these cells will not be used for input or by other worksheet formulas. Your **INPUTS** section should look like the one shown in Figure C-8.

	A	B	C	D
7	INPUTS	2005	2006	2007
8	ECONOMIC OUTLOOK (O = OPTIMISTIC; P=PESSIMISTIC)	NA	O	NA
9	PURCHASE-PRICE OUTLOOK (U = UP; D = DOWN)	NA	U	NA

Figure C-8 Entering two input values

Recall that cell addresses in the **CALCULATIONS** section are already referred to in formulas in the **INCOME STATEMENT AND CASH FLOW STATEMENT** section. The next step is to enter formulas for these calculations. Before doing that, format NUMBER OF UNITS SOLD IN A DAY and NUMBER OF UNITS SOLD IN A YEAR for zero decimals, and format SELLING PRICE PER UNIT and COST OF GOODS SOLD PER UNIT for two decimals.

The easiest formula in the **CALCULATIONS** section is the NUMBER OF UNITS SOLD IN A YEAR, which is the calculated NUMBER OF UNITS SOLD IN A DAY (in C16) times the NUMBER OF BUSINESS DAYS (in C5). In C19, enter =C5*C16, as shown in Figure C-9.

C19 fx =C5*C16

	A	B	C	D
1	TUTORIAL EXERCISE			
2				
3	CONSTANTS	2005	2006	2007
4	TAX RATE	NA	0.33	0.35
5	NUMBER OF BUSINESS DAYS	NA	300	300
6				
7	INPUTS	2005	2006	2007
8	ECONOMIC OUTLOOK (O = OPTIMISTIC; P=PESSIMISTIC)	NA	O	NA
9	PURCHASE-PRICE OUTLOOK (U = UP; D = DOWN)	NA	U	NA
10				
11	SUMMARY OF KEY RESULTS	2005	2006	2007
12	NET INCOME AFTER TAXES	NA	0	0
13	END-OF-THE-YEAR CASH ON HAND	NA	10000	10000
14				
15	CALCULATIONS	2005	2006	2007
16	NUMBER OF UNITS SOLD IN A DAY	1000		
17	SELLING PRICE PER UNIT	7.00		
18	COST OF GOODS SOLD PER UNIT	3.00		
19	NUMBER OF UNITS SOLD IN A YEAR	NA	0	

Figure C-9 Entering the formula to compute 2006 NUMBER OF UNITS SOLD IN A YEAR

Copy the formula to cell D19, for year 2007.

Assume that if the ECONOMIC OUTLOOK is OPTIMISTIC, the 2006 NUMBER OF UNITS SOLD IN A DAY will be 6% more than that in 2005; in 2007, they will be 6% more than that in 2006. Also assume that if the ECONOMIC OUTLOOK is PESSIMISTIC, the NUMBER OF UNITS SOLD IN A DAY in 2006 will be 1% less than those sold in 2005; in

2007, they will be 1% less than those sold in 2006. An =IF() statement is needed in C16 to express this idea:

IF(economy variable = OPTIMISTIC,
then NUMBER OF UNITS SOLD IN A DAY will go UP 6%,
else NUMBER OF UNITS SOLD IN A DAY will go DOWN 1%)

Substituting cell addresses:

=IF(C8 = "O", 1.06 * B16, .99 * B16)

In Excel, quotation marks denote labels. The input is a one-letter label. So, the quotation marks around the **'O'** are needed. You should also note that multiplying by 1.06 results in a 6% rise, whereas multiplying by .99 results in a 1% decrease.

Enter the entire =IF formula into cell C16, as shown in Figure C-10. Absolute addressing is needed (C8) because the address is in a formula that gets copied, *and* you do not want this cell reference to change (to D8, which has the value "NA") when you copy the formula to the right. Absolute addressing maintains the C8 reference when the formula is copied. Copy the formula in C16 to D16 for 2007.

C16 fx =IF(C8="O",1.06*B16,0.99*B16)

	A	B	C	D
15	CALCULATIONS	2005	2006	2007
16	NUMBER OF UNITS SOLD IN A DAY	1000	1060	
17	SELLING PRICE PER UNIT	7.00		
18	COST OF GOODS SOLD PER UNIT	3.00		
19	NUMBER OF UNITS SOLD IN A YEAR	NA	318000	0

Figure C-10 Entering the formula to compute 2006 NUMBER OF UNITS SOLD IN A DAY

The SELLING PRICE PER UNIT is also a function of the ECONOMIC OUTLOOK. The two-part rule is (assume) as follows:

- If the ECONOMIC OUTLOOK is OPTIMISTIC, the SELLING PRICE PER UNIT in 2006 will be 1.07 times that of 2005; in 2007 it will be 1.07 times that of 2006.
- On the other hand, if the ECONOMIC OUTLOOK is PESSIMISTIC, the SELLING PRICE PER UNIT in 2006 and 2007 will equal the per-unit price in 2005 (that is, the price will not change).

Test your understanding of the selling price calculation by figuring out what the formula should be for cell C17. Enter it and copy to the right. You will need to use absolute addressing. (Can you see why?)

The COST OF GOODS SOLD PER UNIT is a function of the PURCHASE-PRICE OUTLOOK. The two-part rule is (assume) as follows:

- If the PURCHASE-PRICE OUTLOOK is UP ("U"), COST OF GOODS SOLD PER UNIT in 2006 will be 1.25 times that of year 2005; in 2007, it will be 1.25 times that of 2006.
- On the other hand, if the PURCHASE-PRICE OUTLOOK is DOWN ("D"), the multiplier in each year will be 1.01.

Again, to test your understanding, figure out what the formula should be in cell C18. Enter it and copy to the right. You will need to use absolute addressing.

Your selling price and cost of goods sold formulas, given OPTIMISTIC and UP input values, should yield the calculated values shown in Figure C-11.

	A	B	C	D
15	**CALCULATIONS**	**2005**	**2006**	**2007**
16	NUMBER OF UNITS SOLD IN A DAY	1000	1060	1124
17	SELLING PRICE PER UNIT	7.00	7.49	8.01
18	COST OF GOODS SOLD PER UNIT	3.00	3.75	4.69
19	NUMBER OF UNITS SOLD IN A YEAR	NA	318000	337080

Figure C-11 Calculated values given OPTIMISTIC and UP input values

Assume that you change the input values to PESSIMISTIC and DOWN. Your formulas should yield the calculated values shown in Figure C-12.

	A	B	C	D
15	**CALCULATIONS**	**2005**	**2006**	**2007**
16	NUMBER OF UNITS SOLD IN A DAY	1000	990	980
17	SELLING PRICE PER UNIT	7.00	7.00	7.00
18	COST OF GOODS SOLD PER UNIT	3.00	3.03	3.06
19	NUMBER OF UNITS SOLD IN A YEAR	NA	297000	294030

Figure C-12 Calculated values given PESSIMISTIC and DOWN input values

That completes the body of your spreadsheet! The values in the **CALCULATIONS** section ripple through the **INCOME STATEMENT AND CASH FLOW STATEMENT** section because the income statement formulas reference the calculations. Assuming inputs of OPTIMISTIC and UP, the income and cash flow numbers should now look like those in Figure C-13.

	A	B	C	D
21	**INCOME STATEMENT AND CASH FLOW STATEMENT**	**2005**	**2006**	**2007**
22	BEGINNING-OF-THE-YEAR CASH ON HAND	NA	10000	806844
23				
24	SALES (REVENUE)	NA	2381820	2701460
25	COST OF GOODS SOLD	NA	1192500	1580063
26	INCOME BEFORE TAXES	NA	1189320	1121398
27	INCOME TAX EXPENSE	NA	392476	392489
28	NET INCOME AFTER TAXES	NA	796844	728909
29				
30	END-OF-THE-YEAR CASH ON HAND (BEGINNING-OF-THE-YEAR CASH, PLUS NET INCOME AFTER TAXES)	10000	806844	1535753

Figure C-13 Completed INCOME STATEMENT AND CASH FLOW STATEMENT section

SCENARIO MANAGER

You are now ready to use the Scenario Manager to capture inputs and results as you play "what if" with the spreadsheet.

Note that there are four possible combinations of input values: O-U (Optimistic-Up), O-D (Optimistic-Down), P-U (Pessimistic-Up), and P-D (Pessimistic-Down). Financial results for each combination will be different. Each combination of input values can be referred to as a "scenario." Excel's Scenario Manager records the results of each combination of input values as a separate scenario and then shows a summary of all scenarios in a separate worksheet. These summary worksheet values can be used as a raw table of numbers, which could be printed or copied into a Word document. The table of data could then be the basis for an Excel chart, which could also be printed or put into a memorandum.

You have four possible scenarios for the economy and the purchase price of goods sold: Optimistic-Up, Optimistic-Down, Pessimistic-Up, and Pessimistic-Down. The four input-value sets produce different financial results. When you use the Scenario Manager, define the four scenarios, then have Excel (1) sequentially run the input values "behind the scenes," and then (2) put the results for each input scenario in a summary sheet.

When you define a scenario, you give the scenario a name and identify the input cells and input values. You do this for each scenario. Then you identify the output cells, so Excel can capture the outputs in a Summary Sheet.

Tutorial C

AT THE KEYBOARD

To start, select Tools—Scenarios. This leads you to a Scenario Manager window. Initially, there are no scenarios defined, and Excel tells you that, as you can see in Figure C-14.

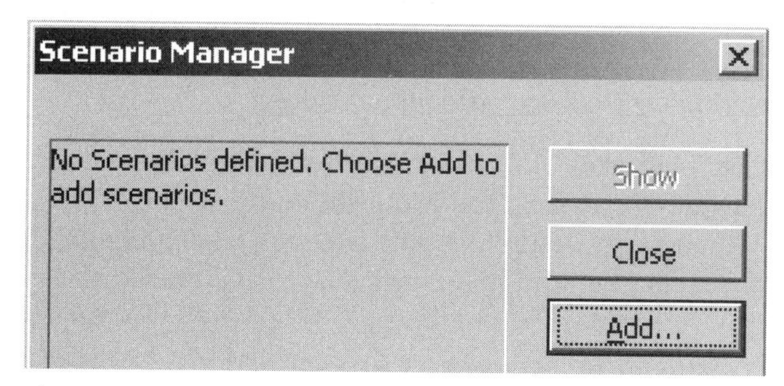

Figure C-14 Initial Scenario Manager window

With this window, you are able to add scenarios, delete them, or change (edit) them. Toward the end of the process, you are also able to create the summary sheet.

NOTE

When working with this window and its successors, do **not** hit the Enter key to navigate. Use mouse clicks to move from one step to the next.

To continue with defining a scenario: Click the Add button. In the resulting Add Scenario window, give the first scenario a name: OPT-UP. Then type in the input cells in the Changing cells window, here, C8:C9. (*Note*: C8 and C9 are contiguous input cells. Non-contiguous input cell ranges can be separated by a comma.) Excel may add dollar signs to the cell address—do not be concerned about this. The window should look like the one shown in Figure C-15.

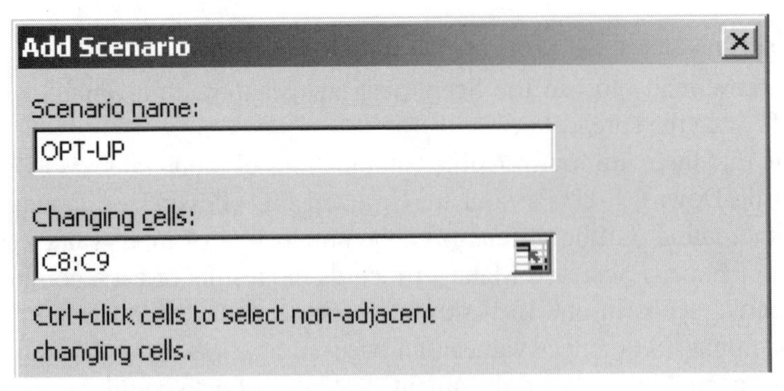

Figure C-15 Entering OPT-UP as a scenario

Now click OK. This moves you to the Scenario Values window. Here you indicate what the INPUT *values* will be for the scenario. The values in the cells *currently in* the spreadsheet will be displayed. They might—or might not—be correct for the scenario you are defining. For the OPT-UP scenario, an O and a U would need to be entered, if not the current values. Enter those values if need be, as shown in Figure C-16.

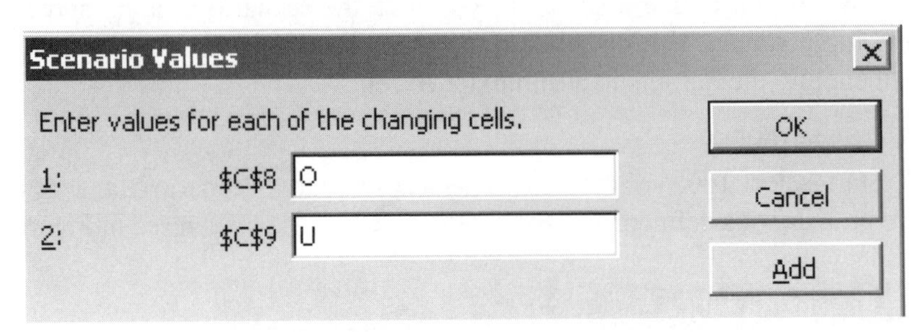

Figure C-16 Entering OPT-UP scenario input values

Now click OK. This takes you back to the Scenario Manager window. You are now able to enter the other three scenarios, following the same steps. Do that now! Enter the OPT-DOWN, PESS-UP, and PESS-DOWN scenarios, plus related input values. After all that, you should see that the names and the changing cells for the four scenarios have been entered, as in Figure C-17.

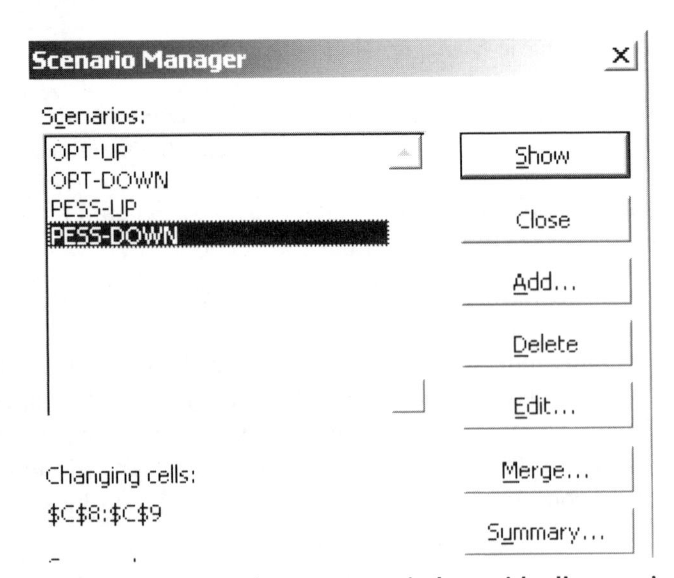

Figure C-17 Scenario Manager window with all scenarios entered

You are now able to create a summary sheet that shows the results of running the four scenarios. Click the Summary button. You'll get the Scenario Summary window. You must tell Excel what the output cell addresses are—these will be the same for all four scenarios. (The output *values* change in those output cells as input values are changed, but the addresses of the output cells do not change.)

Assume that you are primarily interested in the results that have accrued at the end of the two-year period. These are your two 2007 **SUMMARY OF KEY RESULTS** section cells for NET INCOME AFTER TAXES and END-OF-THE-YEAR CASH ON HAND (D12 and D13). Type these addresses into the window's input area, as shown in Figure C-18. (*Note*: If result cells are non-contiguous, the address ranges can be entered, separated by a comma.)

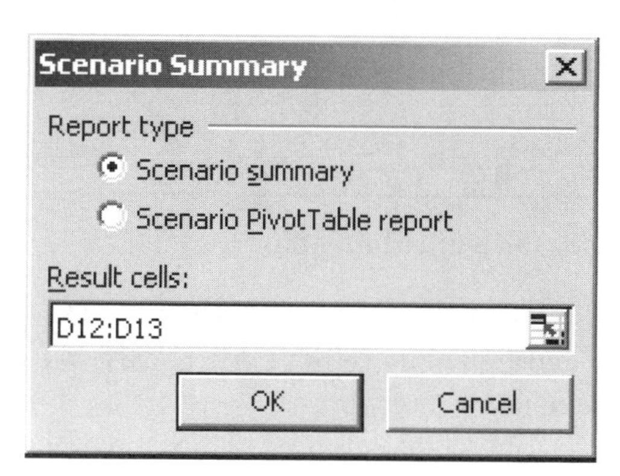

Figure C-18 Entering Result cells addresses in Scenario Summary window

Then click OK. Excel runs each set of inputs and collects results as it goes. (You do not see this happening on the screen.) Excel makes a *new* sheet, titled the Scenario Summary (denoted by the sheet's lower tab), and takes you there, as shown in Figure C-19.

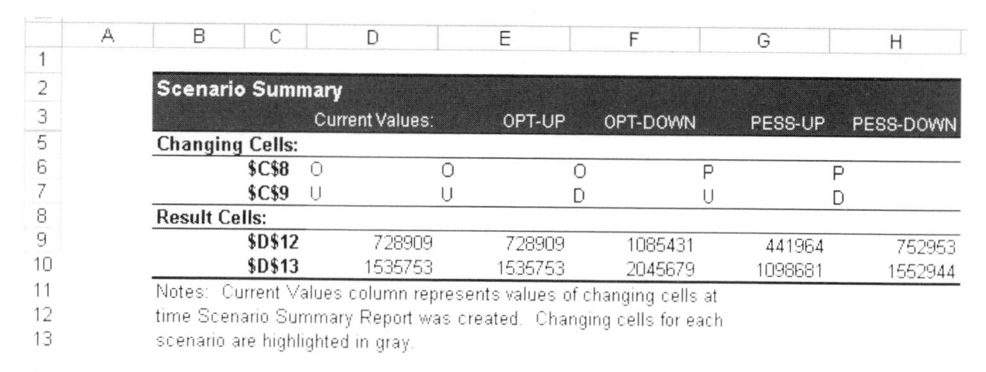

	Current Values:	OPT-UP	OPT-DOWN	PESS-UP	PESS-DOWN
Scenario Summary					
Changing Cells:					
C8	O	O	O	P	P
C9	U	U	D	U	D
Result Cells:					
D12	728909	728909	1085431	441964	752953
D13	1535753	1535753	2045679	1098681	1552944

Notes: Current Values column represents values of changing cells at time Scenario Summary Report was created. Changing cells for each scenario are highlighted in gray.

Figure C-19 Scenario Summary sheet created by Scenario Manager

One somewhat annoying visual element is that the Current Values in the spreadsheet itself are given an output column. This duplicates one of the four defined scenarios. You can delete that extra column by (1) clicking on its column designator letter (here, column D), and (2) clicking Edit—Delete.

Do *not* select Edit—Delete *Sheet*!

Another annoyance is that Column A goes unused. You can click and delete it in the same way to move everything over to the left. This should make columns of data easier to see on the screen, without scrolling. You can also (1) edit cell values to make the results more clear, (2) enter words for the cell addresses, (3) use Alt—Enter to break long headings, if need be, (4) center values using the Format—Cells—Alignment tab menu option, and (5) show data in Currency format, Using the Format—Cells—Number tab menu option.

When you're done, your summary sheet should resemble the one shown in Figure C-20.

	A	B	C	D	E	F
1	**Scenario Summary**					
2			OPT-UP	OPT-DOWN	PESS-UP	PESS-DOWN
4	**Changing Cells:**					
5	**ECONOMIC OUTLOOK**	**C8**	O	O	P	P
6	**PURCHASE PRICE OUTLOOK**	**C9**	U	D	U	D
7	**Result Cells (2007):**					
8	**NET INCOME AFTER TAXES**	**D12**	$728,909	$1,085,431	$441,964	$752,953
9	**END-OF-THE-YEAR CASH ON HAND**	**D13**	$1,535,753	$2,045,679	$1,098,681	$1,552,944

Figure C-20 Scenario Summary sheet after formatting

Note that column C shows the OPTIMISTIC-UP case. NET INCOME AFTER TAXES in that scenario is $728,909, and END-OF-THE-YEAR CASH ON HAND is $1,535,753. Columns D, E, and F show the other scenario results.

Here is an important postscript to this exercise: DSS spreadsheets are used to guide decision-making. This means that the spreadsheet's results must be interpreted in some way. Let's practice with the results shown in Figure C-20. With that data, what combination of year 2007 NET INCOME AFTER TAXES and END-OF-THE-YEAR CASH ON HAND would be best?

Clearly, O-D is the best result, right? It yields the highest income and highest cash. What is the worst combination? P-U, right? It yields the lowest income and lowest cash.

Results are not always this easy to interpret, but the analytical method is the same. You have a complex situation that you cannot understand very well without software assistance. You build a model of the situation in the spreadsheet, enter the inputs, collect the results, and then interpret the results to guide decision-making.

Summary Sheets

When you do Scenario Manager spreadsheet case studies, you'll need to manipulate Summary Sheets and their data. Let's look at some of these operations.

Rerunning the Scenario Manager

To rerun the Scenario Manager, click the Summary button in the Scenario Manager dialog box and then click the OK button. This makes another summary sheet. It does not overwrite a prior one.

Deleting Unwanted Scenario Manager Summary Sheets

Suppose that you want to delete a Summary sheet. With the Summary sheet on the screen, select Edit—Delete Sheet. You will be asked if you really mean it. If so, click to remove, or else cancel out.

Charting Summary Sheet Data

The Summary sheet results can be conveniently charted using the Chart Wizard. Charting Excel data is discussed in Tutorial E.

Copying Summary Sheet Data to the Clipboard

You may want to put the summary sheet data into the Clipboard to use later in a Word document. To do that, use the following steps:

1. Highlight the data range.
2. Use **Edit—Copy** to put the graphic into the Clipboard.
3. Assuming that you want to exit Excel, select **File—Save**, **File—Close**, **File—Exit** Excel. (You may be asked whether you want to leave your data in the Clipboard—you do want to.)
4. Open your Word document.
5. Put your cursor where you want the upper-left part of the graphic to be positioned.
6. Select **Edit—Paste**.

Practice Using Scenario Manager

Suppose that you have an uncle who works for a large company. He has a good job and makes a decent salary ($80,000 a year, currently). He can retire from his company in 2012, when he will be 65. He would start drawing his pension then.

However, the company has an "early out" plan. Under this plan, the company asks employees to quit (called "pre-retirement"). The company then pays those employees a bonus in the year they quit and each year thereafter, up to the official retirement date, which is through the year 2011 for your uncle. Then, employees start to receive their actual pension—in your uncle's case, in 2012. This "early out" program would let your uncle leave the company before 2012. Until then, he could find a part-time hourly-wage job to make ends meet and then leave the workforce entirely in 2012.

The opportunity to leave early is open through 2011. Your uncle could stay with the company in 2006, then pre-retire any time in the years 2007 to 2011, getting the "early out" bonuses in those years. (Of course, if he retires in 2007, he would lose the 2006 bonus, and so on, all the way through 2011.

Another factor in your uncle's thinking is whether to continue belonging to his country club. He likes the club, but it is a real cash drain. The "early out" decision can be looked at each year, but the country club membership decision must be made now—if he does not withdraw in 2006, then he says he will stay a member (and incur costs) through 2011.

Your uncle has called you in to make a Scenario Manager spreadsheet model of his situation. Your spreadsheet would let him play "what if" with the pre-retirement and country club possibilities to see projected 2006-2011 personal finance results. He wants to know what "cash on hand" will be available for each year in the period with each scenario.

Complete the spreadsheet for your uncle. Your **SUMMARY OF KEY RESULTS**, **CALCULATIONS**, and **INCOME STATEMENT AND CASH FLOW STATEMENT** section cells must show values *by cell formula*. That is, in those areas, do not hard-code amounts. In any of your formulas, do not use the address of a cell if its contents are "NA." Set up your spreadsheet skeleton as shown in the figures that follow. Name your spreadsheet **UNCLE.xls**.

CONSTANTS Section

Your spreadsheet should have the constants shown in Figure C-21. An explanation of line items follows the figure.

	A	B	C	D	E	F	G	H
1	**YOUR UNCLE'S EARLY RETIREMENT DECISION**							
2	**CONSTANTS**	**2005**	**2006**	**2007**	**2008**	**2009**	**2010**	**2011**
3	CURRENT SALARY	80000	**NA**	**NA**	**NA**	**NA**	**NA**	**NA**
4	SALARY INCREASE FACTOR	**NA**	0.03	0.03	0.02	0.02	0.01	0.01
5	PART TIME WAGES EXPECTED	**NA**	10000	10200	10500	10800	11400	12000
6	BUY OUT AMOUNT	**NA**	30000	25000	20000	15000	5000	0
7	COST OF LIVING (NOT RETIRED)	**NA**	41000	42000	43000	44000	45000	46000
8	COUNTRY CLUB DUES	**NA**	12000	13000	14000	15000	16000	17000

Figure C-21 CONSTANTS section values

- SALARY INCREASE FACTOR: Your uncle's salary at the end of 2005 will be $80,000. As you can see, raises are expected in each year—for example, a 3% raise is expected in 2006. If he does not retire, he would get his salary and the small raise in a year.
- PART-TIME WAGES EXPECTED: Your uncle has estimated his part-time wages if he were retired and working part time in the 2006-2011 period.
- BUY OUT AMOUNT: The company's pre-retirement "buy out" plan amounts are shown. If your uncle retires in 2006, he gets $30,000, $25,000, $20,000, $15,000, $5,000, and zero in the years 2006 to 2011, respectively. If he leaves in 2006, he gives up the $30,000 2006 payment, but would get $25,000, $20,000, $15,000, $5,000, and zero in the years 2007 to 2011, respectively.
- COST OF LIVING: Your uncle has estimated how much cash he needs to meet his living expenses, assuming that he continues to work for the company. His cost of living would be $41,000 in 2006, increasing each year thereafter.
- COUNTRY CLUB DUES: Country club dues are $12,000 for 2006. They increase each year thereafter.

INPUTS Section

Your spreadsheet should have the inputs shown in Figure C-22. An explanation of line items follows the figure.

	A	B	C	D	E	F	G	H
10	**INPUTS**	**2005**	**2006**	**2007**	**2008**	**2009**	**2010**	**2011**
11	RETIRED [R] or WORKING [W]	**NA**						
12	STAY IN CLUB? [Y] OR [N]	**NA**		**NA**	**NA**	**NA**	**NA**	**NA**

Figure C-22 INPUTS section

- RETIRED OR WORKING: Enter an "R" if your uncle retires in a year, or a "W" if he is still working. If he is working through 2011, the pattern **WWWWWW** should be entered. If his retirement is in 2006, the pattern **RRRRRR** should be entered. If he works for three years and then retires in 2009, the pattern **WWWRRR** should be entered.
- STAY IN CLUB?: If your uncle stays in the club in 2006–2011, a "**Y**" should be entered. If your uncle is leaving the club in 2006, the letter "**N**" should be entered. The decision applies to all years.

SUMMARY OF KEY RESULTS Section

Your spreadsheet should show the results in Figure C-23.

	A	B	C	D	E	F	G	H
14	SUMMARY OF KEY RESULTS	2005	2006	2007	2008	2009	2010	2011
15	END-OF-THE-YEAR CASH ON HAND	NA						

Figure C-23 SUMMARY OF KEY RESULTS section

Each year's END-OF-THE-YEAR CASH ON HAND value is echoed from cells in the spreadsheet body.

CALCULATIONS Section

Your spreadsheet should calculate, by formula, the values shown in Figure C-24. Calculated amounts are used later in the spreadsheet. An explanation of line items follows the figure.

	A	B	C	D	E	F	G	H
17	CALCULATIONS	2005	2006	2007	2008	2009	2010	2011
18	TAX RATE	NA						
19	COST OF LIVING	NA						
20	YEARLY SALARY OR WAGES	80000						
21	COUNTRY CLUB DUES PAID	NA						

Figure C-24 CALCULATIONS section

- TAX RATE: Your uncle's tax rate depends on whether he is retired. Retired people have lower overall tax rates. If he retires in a year, your uncle's rate is expected to be 15% of income before taxes. In a year in which he works full time, the rate will be 30%.
- COST OF LIVING: In any year that your uncle continues to work for the company, his cost of living is what is shown in COST OF LIVING (NOT RETIRED) in the **CONSTANTS** section in Figure C-21. But if he chooses to retire, his cost of living is $15,000 less than the amount shown in the figure.
- YEARLY SALARY OR WAGES: If your uncle keeps working, his salary increases each year. The year-to-year percentage increases are shown in the **CONSTANTS** section. Thus, salary earned in 2006 would be more than that earned in 2005, salary earned in 2007 would be more than that earned in 2006, and so on. If your uncle retires in a certain year, he will make the part-time wages shown in the **CONSTANTS** section.
- COUNTRY CLUB DUES PAID: If your uncle leaves the club, the dues are zero each year; otherwise, the dues are as shown in the **CONSTANTS** section.

The INCOME STATEMENT AND CASH FLOW STATEMENT Section

This section begins with the cash on hand at the beginning of the year. This is followed by the income statement, concluding with the calculation of cash on hand at the end of the year. The format is shown in Figure C-25. An explanation of line items follows the figure.

	A	B	C	D	E	F	G	H
23	**INCOME STATEMENT AND CASH FLOW STATEMENT**	**2005**	**2006**	**2007**	**2008**	**2009**	**2010**	**2011**
24	BEGINNING-OF-THE-YEAR CASH ON HAND	**NA**						
25								
26	SALARY OR WAGES	**NA**						
27	BUY OUT INCOME	**NA**						
28	TOTAL CASH INFLOW	**NA**						
29	COUNTRY CLUB DUES PAID	**NA**						
30	COST OF LIVING	**NA**						
31	TOTAL COSTS	**NA**						
32	INCOME BEFORE TAXES	**NA**						
33	INCOME TAX EXPENSE	**NA**						
34	NET INCOME AFTER TAXES	**NA**						
35								
36	END-OF-THE-YEAR CASH ON HAND (BEGINNING-OF-THE-YEAR CASH, PLUS NET INCOME AFTER TAXES)	30000						

Figure C-25 INCOME STATEMENT AND CASH FLOW STATEMENT section

- BEGINNING-OF-THE-YEAR CASH ON HAND: This is the END-OF-THE-YEAR CASH ON HAND at the end of the prior year.
- SALARY OR WAGES: This is a yearly calculation.
- BUY OUT INCOME: This is the year's "buy out" amount, if your uncle retires in the year.
- TOTAL CASH INFLOW: This is the sum of salary or part-time wages and "buy out" amounts.
- COUNTRY CLUB DUES PAID: This is a calculated amount.
- COST OF LIVING: This is a calculated amount.
- TOTAL COSTS: These outflows are the sum of the COST OF LIVING and COUNTRY CLUB DUES PAID.
- INCOME BEFORE TAXES: This amount is the TOTAL CASH INFLOW less TOTAL COSTS (outflows).
- INCOME TAX EXPENSE: This amount is zero if INCOME BEFORE TAXES is zero or less; otherwise, the calculated tax rate is applied to the INCOME BEFORE TAXES.
- NET INCOME AFTER TAXES: This is INCOME BEFORE TAXES, less TAX EXPENSE.
- END-OF-THE-YEAR CASH ON HAND: This is the BEGINNING-OF-THE-YEAR CASH plus the year's NET INCOME AFTER TAXES.

Scenario Manager Analysis

Set up the Scenario Manager and create a Scenario Summary sheet. Your uncle wants to look at the following four possibilities:

- Retire in 2006, staying in the club ("Loaf-In")
- Retire in 2006, leaving the club ("Loaf-Out")
- Work three more years, retire in 2009, staying in the club ("Delay-In")
- Work three more years, retire in 2009, leaving the club ("Delay-Out")

You can enter non-contiguous cell ranges as follows: C20..F20, C21, C22 (cell addresses are examples). The output cell should be the 2011 (only) END-OF-THE-YEAR CASH ON HAND cell.

Your uncle will choose the option that yields the highest 2011 END-OF-THE-YEAR CASH ON HAND. You must look at your Scenario Summary sheet to see which strategy yields the highest amount.

To check your work, you should attain the values shown in Figure C-26. (You can use the labels Excel gives you in the left-most column or change the labels, as was done in Figure C-26.)

	A	B	C	D	E	F
1	Scenario Summary					
2			LOAF-IN	LOAF-OUT	DELAY-IN	DELAY-OUT
4	Changing Cells:					
5	RETIRE OR WORK, 2006	C11	R	R	W	W
6	RETIRE OR WORK, 2007	D11	R	R	W	W
7	RETIRE OR WORK, 2008	E11	R	R	W	W
8	RETIRE OR WORK, 2009	F11	R	R	R	R
9	RETIRE OR WORK, 2010	G11	R	R	R	R
10	RETIRE OR WORK, 2011	H11	R	R	R	R
11	IN CLUB 2006-2011?	C12	Y	N	Y	N
12	Result Cells (2011):					
13	END-OF-THE-YEAR CASH ON HAND	H15	-$68,400	$15,195	$8,389	$83,689

Figure C-26 Scenario Summary

Review of Excel Basics

In this section, you'll begin by reviewing how to perform some basic operations. Then, you'll work through some further cash flow calculations. Reading and working through this section will help you to do the spreadsheet cases in this book.

Basic Operations

In this section, you'll review the following topics: formatting cells, showing Excel cell formulas, understanding circular references, using the And and the Or functions in IF statements, and using nested IF statements.

Formatting Cells

You may have noticed that some data in this tutorial's first spreadsheet was centered in the cells. Here is how to perform that operation:

1. Highlight the cell range to format.
2. Select the **Format** menu option.
3. Select **Cells—Alignment**.
4. Choose **Center** for both **Horizontal** and **Vertical**.
5. Select **OK**.

It is also possible to put a border around cells. This treatment might be desirable for highlighting **INPUTS** section cells. To perform this operation:

1. Select **Format—Cells—Border—Outline**.
2. Choose the outline **Style** you want.
3. Select **OK**.

You can format numerical values for Currency format by selecting:

Format—Cells—Number—Currency.

You can format numerical values for decimal places using this procedure:

1. Select **Format—Cells—Number** tab—**Number**.
2. Select the desired number of decimal places.

Showing Excel Cell Formulas

If you want to see Excel cell formulas, follow this procedure:

1. Press the **Ctrl** key and the "back-quote" key (`) at the same time. (The back-quote orients from Seattle to Miami—on most keyboards, it is next to the exclamation-point key and shares the key with the tilde diacritic mark.)
2. To restore, press the **Ctrl** key and (`) back-quote key again.

Understanding a Circular Reference

A formula has a circular reference if the *formula refers to the cell that the formula is already in*. Excel cannot evaluate such a formula, because the value of the cell is not yet known—but to do that evaluation, the value in the cell must be known! The reasoning is circular, hence the term "circular reference." Excel will point out circular references, if any exist, when you choose Open for a spreadsheet. Excel will also point out circular references as you insert them during development. Excel will be demonstrative about this by opening at least one Help window and by drawing arrows between cells involved in the offending formula. You can close the windows, but that will not fix the situation. You *must* fix the formula that has the circular reference if you want the spreadsheet to give you accurate results.

Here is an example. Suppose that the formula in cell C18 is =C18 – C17. Excel tries to evaluate the formula in order to put a value on the screen in cell C18. To do that, Excel must know the value in cell C18—but that is what it is trying to do in the first place. Can you see the circularity?

Using the "And" Function and the "Or" Function in =IF Statements

An =IF() statement has the following syntax:

=IF(test condition, result if test is True, result if test is False)

The test conditions in this tutorial's =IF statements tested only one cell's value. A test condition could test more than one cell's values.

Here is an example from this tutorial's first spreadsheet. In that example, selling price was a function of the economy. Assume, for the sake of illustration, that year 2005's selling price per unit depends on the economy *and* the purchase price outlook. If the economic outlook is optimistic *and* the company's purchase price outlook is down, then the selling price will be 1.10 times the prior year's price. Assume that in all other cases, the selling price will be 1.03 times the prior year's price. The first test requires two things to be true *at the same time*: C8 = "O" *AND* C9 = "D." So, the AND() function would be needed. The code in cell C17 would be as follows:

=IF(AND(C8 = "O", C9 = "D"), 1.10 * B17, 1.03 * B17)

On the other hand, the test might be this: If the economic outlook is optimistic *or* the purchase price outlook is down, then the selling price will be 1.10 times the prior year's price. Assume that in all other cases, the selling price will be 1.03 times the prior year's price. The first test requires *only one of* two things to be true: C8 = "O" *or* C9 = "D". Thus, the OR() function would be needed. The code in cell C17 would be:

=IF(OR(C8 = "O", C9 = "D"), 1.10 * B17, 1.03 * B17)

Using IF() Statements Inside IF() Statements

An =IF() statement has this syntax:

=IF(test condition, result if test is True, result if test is False)

In the examples shown thus far, only two courses of action were possible, so only one test has been needed in the =IF() statement. There can be more courses of action than two, however, and this requires that the "result if test is False" clause needs to show further testing. Let's look at an example.

Assume again that the 2006 selling price per unit depends on the economy and the purchase price outlook. Here is the logic: (1) If the economic outlook is optimistic *and* the purchase price outlook is down, then the selling price will be 1.10 times the prior year's price. (2) If the economic outlook is optimistic *and* the purchase price outlook is up, then the selling price will be 1.07 times the prior year's price. (3) In all other cases, the selling price will be 1.03 times the prior year's price. The code in cell C17 would be:

=IF(AND(C8 = "O", C9 = "D"), 1.10 * B17,
IF(AND(C8 = "O", C9 = "U"), 1.07 * B17, 1.03 * B17))

The first =IF() tests to see if the economic outlook is optimistic and the purchase price outlook is down. If not, further testing is needed to see whether the economic outlook is optimistic and the purchase price outlook is up, or whether some other situation prevails.

NOTE

Be sure to note the following:

- The line is broken in the previous example because the page is not wide enough, but in Excel, the formula would appear on one line.
- The embedded "IF" is not preceded by an equals sign.

Example: Borrowing and Repayment of Debt

The Scenario Manager cases require you to account for money that the company borrows or repays. Borrowing and repayment calculations are discussed next. At times you are asked to think about a question and fill in the answers. Correct responses are at the end of this section.

To do the Scenario Manager cases, you must assume two things about a company's borrowing and repayment of debt. First, assume that the company wants to have a certain minimum cash level at the end of a year (and thus to start the next year). Assume that a bank will provide a loan to reach the minimum cash level if year-end cash falls short of that level.

Here are some numerical examples to test your understanding. Assume that NCP stands for "net cash position" and equals beginning-of-the-year cash plus net income after taxes for the year. The NCP is the cash available at year end, before any borrowing or repayment. Compute the amounts to borrow in the three examples in Figure C-27.

Example	NCP	Minimum Cash Required	Amount to Borrow
1	50,000	10,000	?
2	8,000	10,000	?
3	−20,000	10,000	?

Figure C-27 Examples of borrowing

Assume that a company would use its excess cash at year end to pay off as much debt as possible, without going below the minimum-cash threshold. "Excess cash" would be the NCP *less* the minimum cash required on hand—amounts over the minimum are available to repay any debt.

In the examples shown in Figure C-28, compute excess cash and then compute the amount to repay. You may also want to compute ending cash after repayments as well, to aid your understanding.

Example	NCP	Minimum Cash Required	Beginning-of-the-Year Debt	Repay	Ending Cash
1	12,000	10,000	4,000	?	?
2	12,000	10,000	10,000	?	?
3	20,000	10,000	10,000	?	?
4	20,000	10,000	0	?	?
5	60,000	10,000	40,000	?	?
6	–20,000	10,000	10,000	?	?

Figure C-28 Examples of repayment

In this section's Scenario Manager cases, your spreadsheet will need two bank financing sections beneath the **INCOME STATEMENT AND CASH FLOW STATEMENT** section:

1. The first section will calculate any needed borrowing or repayment at the year's end to compute year-end cash.
2. The second section will calculate the amount of debt owed at the end of the year, after borrowing or repayment of debt.

The first new section, in effect, extends the end-of-year cash calculation, which was shown in Figure C-13. Previously, the amount equaled cash at the beginning of the year plus the year's net income. Now, the calculation will include cash obtained by borrowing and cash repaid. Figure C-29 shows the structure of the calculation.

	A	B	C	D
30	NET CASH POSITION (NCP) BEFORE BORROWING AND REPAYMENT OF DEBT (BEGINNING-OF-THE-YEAR CASH PLUS NET INCOME AFTER TAXES)	NA		
31	PLUS: BORROWING FROM BANK	NA		
32	LESS: REPAYMENT TO BANK	NA		
33	EQUALS: END-OF-THE-YEAR CASH ON HAND	10000		

Figure C-29 Calculation of END-OF-THE-YEAR CASH ON HAND

The heading in cell A30 was previously END-OF-THE-YEAR CASH ON HAND. But BORROWING increases cash and REPAYMENT OF DEBT decreases cash. So, END-OF-THE-YEAR CASH ON HAND is now computed two rows down (in C33 for year 2006, in the example). The value in row 30 must be a subtotal for the BEGINNING-OF-THE-YEAR CASH ON HAND plus the year's NET INCOME AFTER TAXES. We call that subtotal the NET CASH POSITION (NCP) BEFORE BORROWING AND REPAYMENT OF DEBT.

(*Note*: Previously, the formula in cell C22 for BEGINNING-OF-THE-YEAR CASH ON HAND was =B30. Now, that formula would be =B33. It would be copied to the right, as before, for the next year.)

That second new section would look like what is shown in Figure C-30.

	A	B	C	D
35	DEBT OWED	2005	2006	2007
36	BEGINNING-OF-THE-YEAR DEBT OWED	NA		
37	PLUS: BORROWING FROM BANK	NA		
38	LESS: REPAYMENT TO BANK	NA		
39	EQUALS: END-OF-THE-YEAR DEBT OWED	15000		

Figure C-30 DEBT OWED section

The second new section computes end-of-year debt and is called DEBT OWED. At the end of 2005, $15,000 was owed. END-OF-THE-YEAR DEBT OWED equals the BEGINNING-OF-THE-YEAR DEBT OWED, plus any new BORROWING FROM BANK (which increases debt owed), less any REPAYMENT TO BANK (which reduces it). So, in the example, the formula in cell C39 would be:

=C36 + C37 – C38

Assume that the amounts for BORROWING FROM BANK and REPAYMENT TO BANK are calculated in the first new section. Thus, the formula in cell C37 would be: =C31. The formula in cell C38 would be: =C32. (BEGINNING-OF-THE-YEAR DEBT OWED is equal to the debt owed at the end of the prior year, of course. The formula in cell C36 for BEGINNING-OF-THE-YEAR DEBT OWED would be an echoed formula. *Can you see what it would be*? It's an exercise for you to complete. *Hint*: The debt owed at the beginning of a year equals the debt owed at the end of the prior year.)

Now that you have seen how the borrowing and repayment data is shown, the logic of the borrowing and repayment formulas can be discussed.

Calculation of BORROWING FROM BANK

The logic of this in English is:

If (cash on hand before financing transactions is greater than the minimum cash required, then borrowing is not needed; otherwise, borrow enough to get to the minimum).

Or (a little more precisely):

If (NCP is greater than the minimum cash required, then BORROWING FROM BANK = 0; otherwise, borrow enough to get to the minimum).

Suppose that the desired minimum cash at year end is $10,000, and that value is a constant in your spreadsheet's cell C6. Assume the NCP is shown in your spreadsheet's cell C30. Our formula (getting closer to Excel) would be as follows:

IF(NCP > Minimum Cash, 0; otherwise, borrow enough to get to the minimum).

You have cell addresses that stand for NCP (cell C30) and Minimum Cash (C6). To develop the formula for cell C31, substitute these cell addresses for NCP and Minimum Cash. The harder logic is that for the "otherwise" clause. At this point, you should look ahead to the borrowing answers at the end of this section, Figure C-31. In Example 2, $2,000 had to be borrowed. Which cell was subtracted from which other cell to calculate that amount? Substitute cell addresses in the Excel formula for the year 2005 borrowing formula in cell C31:

```
=IF(            >=       , 0 ,      -        )
```

The Answer is at the end of this section, Figure C-33.

Calculation of REPAYMENT TO BANK

The logic of this in English is:

IF(beginning of year debt = 0, repay 0 because nothing is owed, but
    IF(NCP is less than the min, repay zero, because you must *borrow*, but
        IF(extra cash equals or exceeds the debt, repay the whole debt,
            ELSE (to stay above the min, repay only the extra cash))))

Look at the following formula. Assume the repayment will be in cell C32. Assume debt owed at the beginning of the year is in cell C36, and that minimum cash is in cell C6. Substitute cell addresses for concepts to complete the formula for year 2006 repayment. (Clauses are on different lines because of page width limitations.)

```
=IF(            = 0, 0,
        IF(            <=              , 0,
                IF( (              -        ) >=    ,
                    (              -                  ) ))).
```

The answer is at the end of this section, in Figure C-34.

Answers to Questions About Borrowing and Repayment Calculations

Figures C-31 and C-32 give the answers to the questions posed about borrowing and repayment calculations.

Example	NCP	Minimum Cash Required	Borrow	Comments
1	50,000	10,000	Zero	NCP > Min.
2	8,000	10,000	2,000	Need 2000 to get to Min. 10,000 – 8,000
3	–20,000	10,000	30,000	Need 30000 to get to Min. 10,000 – (–20,000)

Figure C-31 Answers to examples of borrowing

Example	NCP	Minimum Cash Required	Beginning-of-the-Year Debt	Repay	Ending Cash
1	12,000	10,000	4,000	2,000	10,000
2	12,000	10,000	10,000	2,000	10,000
3	20,000	10,000	10,000	10,000	10,000
4	20,000	10,000	0	0	20,000
5	60,000	10,000	40,000	40,000	20,000
6	–20,000	10,000	10,000	NA	NA

Figure C-32 Answers to examples of repayment

Some notes about the repayment calculations shown in Figure C-32 follow.

- In Examples 1 and 2, only $2,000 is available for debt repayment (12,000 – 10,000) to avoid going below the minimum cash.
- In Example 3, cash available for repayment is $10,000 (20,000 – 10,000), so all beginning debt can be repaid, leaving the minimum cash.
- In Example 4, there is no debt owed, so no debt need be repaid.
- In Example 5, cash available for repayment is $50,000 (60,000 – 10,000), so all beginning debt can be repaid, leaving more than the minimum cash.
- In Example 6, no cash is available for repayment. The company must borrow.

Figures C-33 and C-34 show the calculations for borrowing and repayment of debt.

```
=IF( C30 >= C6, 0, C6 – C30)
```

Figure C-33 Calculation of borrowing

```
=IF( C36 = 0, 0, IF( C30 <= C6, 0, IF( (C30 – C6) >= C36, C36, ( C30 – C6) )))
```

Figure C-34 Calculation of repayment

Saving Files After Using Microsoft Excel

As you work, save periodically (**File—Save**). If you want to save to a diskette, choose **Drive A:**. At the end of a session, save your work using this three-step procedure:

1. Save the file, using **File—Save**. If you want to save to a diskette, choose **Drive A:**.
2. Use **File—Close** to tell Windows to close the file. If saving to a diskette, make sure it is still in **Drive A:** when you close. If you violate this rule, you may lose your work!

3. Exit from Excel to Windows by selecting **File—Exit**. In theory, you may exit from Excel back to Windows after you have saved a file (short-cutting the File—Close step), but that is not a recommended shortcut.

Tutorial C

7 CASE

BigState Bank's Call Center Off-Shoring Decision

Decision Support Using Excel

Preview

BigState Bank employs hundreds of employees in their call center, which is located in the United States. Bank management is considering laying off these employees and establishing a call center in Bangalore, India, to take advantage of that country's much lower salary levels. However, there are hidden costs to "off-shoring." In this case, you will use Excel to see whether the financial benefits of the Bangalore move would outweigh the costs.

Preparation

- Review spreadsheet concepts discussed in class and/or in your textbook. This case requires an understanding of these Excel functions: INT(), NPV(), IF(), and SUM().
- Complete any exercises that your instructor assigns.
- Complete any part of Tutorial C that your instructor assigns, or refer to it as necessary.
- Review file-saving procedures for Windows programs. These are discussed in Tutorial C.
- Refer to Tutorial E as necessary.

BACKGROUND

BigState Bank has substantial credit card and banking operations throughout the U.S. The bank is headquartered in a large Midwestern state. The bank has 800 customer representatives, who deal with bank customers over the phone. The representatives' job is to answer credit card queries, collect past-due balances, make "cold calls" to sell new credit cards, answer questions about the bank's Certificates of Deposits and other investments, and so on. The customer representatives work in a large "call center" located in an industrial campus outside the state's capitol.

Call center operations seem to be good; however, BigState Bank thinks that the center is expensive. A call center representative earns about $30,000 a year in salary and benefits, and call center infrastructure costs are high. Many U.S. companies are transferring business operations to "off-shore" countries where employment costs are significantly lower. BigState Bank wants to evaluate the option of shifting their call center to Bangalore, India, to take advantage of that country's lower wage rates. In addition, unlike U.S. call center workers, call center workers in India are usually college graduates. Thus, BigState Bank expects Indian call center workers to provide a higher level of service.

If the move is made, all U.S. call center employees would be laid off, and the call center real estate would be devoted to another purpose. An Indian firm, Bangalore InfoSystems (BI), would be hired as a contractor to run the bank's call center operations. BI would arrange for a site, hire and train the staff, and run the center's day-to-day operations. BigState Bank would pay BI in three ways:

1. A flat fee per employee (this fee would cover salaries and infrastructure costs)
2. A fee to handle the security check on each new employee
3. A fee for training each new employee

Given these payments, BI would run the call center while trying to make a profit. The agreement would be hammered out in 2005 and would be in effect for the five years (2006–2010).

BigState Bank expects these benefits from the operation:

- Lower wage rates and a high quality of work
- Assets that can be re-deployed in the business, including existing real estate and current call center supervisors
- A refocus on banking operations, their "core" business, rather than spending time and effort running a call center
- Cash that would be freed up by foregoing telecommunication, computing, and other infrastructure costs

An Indian call center worker is paid about 80% less than a U.S. call center worker, and the Indians are highly skilled. With 800 such workers, the decision, at first, looks easy. However, there are a number of things to consider. These factors are discussed next.

Factors to Consider

Selecting a Contractor

Evaluating competing contractors in the foreign country is difficult and time consuming. Bids must be obtained and evaluated, Indian sites must be inspected, and so forth. In 2005, BI was finally identified as the best contractor.

Legal and Technical Consulting

A U.S. company needs foreign representation in a foreign country, and BigState Bank has had to hire a variety of "in country" representatives and consultants. These are described next.

1. To begin, BigState Bank retained an Indian law firm, and their services have been expensive. Such services will continue to be expensive during the life of the contract with BI.
2. Before selecting BI as a contractor, BigState Bank hired an Indian technical consulting firm to evaluate BI's ability as a contactor. This review covered a wide range of issues: BI's telecommunications contracts, its plan for air conditioning the call center site, background and security checks on management and supervisors, and its plan for disaster recovery—what would happen if the Bangalore center was shut down due to a catastrophe? There are always fresh technology and management security issues, and during the contract, BigState Bank would continue to use the technology consultant.
3. Managing a foreign contract requires a good deal of executive-to-executive contact. Much of this can be done by e-mail and video conferencing. However, frequent visits to Bangalore will be needed, and this executive business travel is also considered a consulting expense.
4. BI would handle the actual recruiting of new employees, but BigState Bank wants potential workers in India to know about the bank. Therefore, BigState Bank would hire a public relations (PR) firm in India to burnish its image, and the bank would retain the firm during the contract years.

Liaison Employees

Companies that have outsourced labor in India have noticed that day-to-day communication is more successful if some representatives are stationed with their contractor. Thus, BigState Bank would send two of its people to Bangalore to work with BI. BI would send two of its employees to work with BigState Bank in the United States. Paying for an employee to live and work in a foreign country is expensive but necessary.

Background Checks for New Employees

As a security measure, the bank requires that all new call center employees have a background check. Background checks would also be required for all new non-supervisory employees in India. In the U.S., background checks are inexpensive because so much information about people is readily available and recorded in digital format. In India, however, background information is difficult to obtain and, thus, background checks would be time consuming and expensive.

Training Costs

It costs about $1,000 to train a call center worker in the U.S. Training will cost more in India. As with their U.S. counterparts, Indian workers must be trained on technical aspects of banking; in addition, Indian workers must also be trained to *speak* like Americans. When a customer from New York calls about a credit card error, he does not want to be distracted or confused by someone with a foreign accent. For the same reason, Indian workers must be trained in U.S. cultural norms.

Indian Labor Market

As U.S. demand for off-shore operations increases, so does the demand for qualified foreign labor. Wage rates are expected to escalate in the future, and this will affect payments to BI. In addition, turnover is expected to increase as workers change jobs, seeking higher pay. During the contract period, turnover rates are expected to be higher than in the U.S.

Indian technical schools are expected to turn out more and more qualified people. The level of salary inflation in India is a key unknown in the decision, as is the employee turnover rate.

Exchange Rate Problems

During the life of the contract, BI management wants to be paid in rupees, not in dollars. Of course, BigState Bank has dollars and operates in a dollar-based economy. Thus, BigState Bank must exchange dollars for rupees to pay BI.

At the time of the contract with BI, $1.00 buys 45.85 rupees. (Stated another way, a rupee is worth $.02181, or 1/45.85). During the life of the contract, all costs will be based on that fixed equivalent between dollars and rupees—even though daily international monetary exchange markets will fluctuate. Thus, all fees owed in rupees during the life of the contract will be computed in this three-step process:

1. The fee is computed first in dollars.
2. The fee will then be converted to rupees at the fixed rate of 45.85.
3. Then, BigState Bank would acquire the needed rupees, at the current international exchange rate (which fluctuates), and send funds in rupees to BI.

Example: BigState Bank owes BI $1,000,000 for some service. Expressed in rupees at the agreed upon fixed rate of 45.85, this equals 45,850,000 rupees. Because international money markets fluctuate, BigState Bank will need to factor in that fluctuation when they buy rupees to send to BI. Suppose the dollar on the international market exchange rate has "strengthened" to 47 rupees to a dollar. BigState Bank must pay BI 45.85 million rupees, but how many dollars are required to pay off the debt? This is the calculation:

45,850,000 rupees / 47 = $975, 532

That is how many dollars would be needed at the bank to satisfy the debt. In effect, BigState Bank has gained $24,468 ($1 million less $975,532) by the strengthening of the dollar versus the rupee in the foreign exchange markets. Fewer dollars are needed to buy the 45.85 million rupees.

Note that U.S. tourists are well aware of this effect. If a tourist books a vacation in Europe and the dollar strengthens by the time the tourist gets there, the tourist's dollars can buy more goods priced in Euros than previously expected.

The exchange rate can go the other way too. Suppose the dollar-to-rupee exchange rate "weakens" to 44. BigState Bank must still pay 45.85 million rupees. How many dollars are required to pay off the debt? The answer is calculated as follows:

45,850,000 rupees / 44 = $1,042,046

In effect, BigState Bank has lost $42,045 by the weakening of the dollar versus the rupee. (Experienced U.S. tourists are all too aware of this possibility as well!)

As you will see, exchange rate fluctuations will be important in calculating four kinds of payments to BI: (1) the cost of call center employees, (2) the cost of liaison employees, (3) the cost of background checks, and (4) the cost of training.

Economic Analysis: Using Net Present Value (NPV)

The BI contract will last five years. BigState Bank uses the Net Present Value (NPV) approach to analyze multi-period investments. This approach is explained next.

Money has a "time value." A dollar today is worth more than a dollar promised to you a year from now. This fact changes the way you must think about long-term investments. To illustrate this point, imagine that you have $6,000 to invest, and that you have an opportunity to make a 10% return on the investment. Suppose that a banker tells you that if you invest your $6,000 with her for one year, you will receive $6,500 a year from now. Should you do it? No. The $6,000 invested at 10% for a year would yield $6,600; the banker's offer will yield only $6,500—a shortfall of $100.

The proper way to look at this is to restate all investment cash flows in terms of today's dollars. Use a discount rate for each interest rate and each future year to reduce ("discount") future dollars to today's dollars. Continuing the example, suppose that the discount rate for 10% one year in the future is equal to 1 / (1 + .10), or .90909. Thus, the $6,500 you receive a year from now should earn 10% interest. However,

$6,500 * .90909 = $5,909

Compared to the $6,000 you started with, the net present value (the NVP) is $91 negative! Figure 7-1 illustrates the discount.

Cash Flows in Today's Dollars		
Cash Out Now	**Year 1 Cash Inflow**	**Difference**
–$6,000	$5,909	–$91

Figure 7-1 NPV Example

Continuing the example, suppose that the banker had offered to pay $6,800 a year hence. In today's dollars, that would be equivalent to $6,800 * .90909, or $6,182. In this case the NPV is a positive $182 (–$6,000 + $6,182.)

There is a discount factor for each year and each assumed investment earnings rate: For example, the discount rate for 10% two years in the future is equal to $1 / (1 + .10)^2$, or .8264. This example shows that the longer you must wait to receive cash, the less it is worth in today's dollars (the more heavily it is discounted).

There is a well-known series of steps for evaluating investments by NPV:

1. Cash is expended in the first year. The outflow is already in today's dollars, so it is not discounted. This is treated as a negative number, as cash that flowed out of the company.
2. In succeeding years, the investment (hopefully) returns more cash to the company than it costs to maintain the investment. Net cash flows (inflows less outflows) in each year should be positive numbers over the investment's useful life.
3. The net cash flows are each discounted back to present-day dollars, using the discount factor for the appropriate rate and year. These discounted flows are added up, thus restating all the years' cash flows in Year 1 dollars.
4. The first year's outflow is a negative number. In effect, including it in the summation subtracts it from the total to give an NPV of all the investment cash flows for the life of the project. If this overall NPV is positive, the investment should be considered. If the overall NPV is negative, the investment should not be made.

The appropriate discount rate for a company is sometimes called the "hurdle rate" because it is the minimum acceptable rate of earnings on a potential investment for that company. BigState Bank's hurdle rate for NPV analyses has been 25%, applied to cash flows before taxes. That is, the company looks for investments that have a positive NPV at 25%.

Thus, if BigState Bank's potential investment shows a positive dollar NPV at 25%, the Chief Information Officer (CIO) will be financially justified in moving the call center to India. If BigState Bank's potential investment shows a negative NPV at 25%, the CIO will not be financially justified in moving the call center to India.

ASSIGNMENT 1 CREATING A SPREADSHEET FOR DECISION SUPPORT

In this assignment, you will produce a spreadsheet that models the business decision. Then, in Assignment 2, you will write a memorandum to the bank's CIO that explains your recommended action. In addition, in Assignment 3, you will be asked to prepare an oral presentation of your analysis and recommendation.

Next, you will create the spreadsheet model of the off-shoring decision. The model is an NPV analysis for the years 2005-2010 (2005 is the year money is invested—for the NPV analysis, 2005 dollars are "today's dollars"). You will be given some hints on how each section should be set up before entering cell formulas. Your spreadsheet should have the sections that follow.

- **CONSTANTS**
- **INPUTS**
- **SUMMARY OF KEY RESULTS**
- **CALCULATIONS**
- **NET CASH FLOW ANALYSIS**

A discussion of each section follows. *The spreadsheet skeleton is available to you, so you need not type in the skeleton if you do not wish to do so.* To access the spreadsheet skeleton, go to your Data files. Select Case 7, then select **OFFSHORE.xls**.

CONSTANTS Section

Your spreadsheet should have the constants shown in Figure 7-2. An explanation of the line items follows the figure.

	A	B	C	D	E	F	G
1	**BIGSTATE BANK'S CALL CENTER OFF-SHORING DECISION**						
2							
3	CONSTANTS	2005	2006	2007	2008	2009	2010
4	U.S. TELECOMM COSTS, PER EMPLOYEE	NA	95	90	86	80	75
5	U.S. COMPUTING COSTS, PER EMPLOYEE	NA	390	380	370	360	350
6	REAL ESTATE COST, PER U.S. EMPLOYEE	NA	1100	1200	1300	1400	1500
7	EMPLOYEE-SUPERVISOR RATIO, U.S.	NA	20	20	20	20	20
8	AVERAGE U.S. EMPLOYEE EMPLOYMENT COST	NA	30000	31000	32000	33000	34000
9	AVERAGE SUPERVISOR EMPLOYMENT COST	NA	52000	53000	54000	55000	56000
10	U.S. EMPLOYEE TURNOVER RATE	NA	0.11	0.12	0.13	0.14	0.15
11	TRAINING COST PER U.S. EMPLOYEE	NA	1100	1200	1300	1400	1500
12	TRAINING COST PER INDIAN EMPLOYEE	NA	1600	1700	1800	1900	2000
13	LEGAL AND TECHNICAL CONSULTING -- INDIA	15000000	6000000	6500000	7000000	7500000	8000000
14	TOTAL NUMBER OF LIAISON EMPLOYEES	NA	4	4	4	4	4
15	COST OF LIAISON EMPLOYEE	NA	200000	210000	220000	230000	240000
16	COST OF EMPLOYEE BUYOUT	NA	2000	1500	1000	500	0
17	COST OF BACKGROUND CHECK -- INDIA	NA	1000	1100	1200	1300	1400
18	COST OF BACKGROUND CHECK -- U.S.	NA	20	25	30	35	40

Figure 7-2 CONSTANTS section

Case 7

- U.S. TELECOM COSTS, PER EMPLOYEE: At the U.S. call center, networking and telecommunication costs would be $95 per non-supervisory employee in 2006. The yearly rate would decline in future years. If the center is off-shored, BigState Bank will not have to pay for telecommunications directly.
- U.S. COMPUTING COSTS, PER EMPLOYEE: At the U.S. call center, the cost of computer support would be $390 per non-supervisory employee in 2006. The yearly rate would decline in future years. If the center is off-shored, BigState Bank will not have to pay for computer support directly.
- REAL ESTATE COST, PER U.S. EMPLOYEE: At the U.S. call center, the cost of real estate would be $1,100 per non-supervisory employee in 2006 (upkeep on the call center building, insurance, etc.). The yearly rate would increase in future years. If the center is off-shored, BigState Bank would be able to use the site for different purposes—this would be considered a benefit of the project.
- EMPLOYEE-TO-SUPERVISOR RATIO, U.S.: For every 20 U.S. call center employees, BigState Bank employs 1 supervisor. If the center is off-shored, supervisors can be used in other areas of the bank—this would be considered a benefit of the project.
- AVERAGE U.S. EMPLOYEE EMPLOYMENT COST: The average cost of a U.S. call center employee will be $30,000 in 2006 for salary and benefits. This rate would increase in future years.
- AVERAGE SUPERVISOR EMPLOYMENT COST: The average cost of a U.S. call center supervisor will be $52,000 in 2006 for salary and benefits. This rate would increase in future years.
- U.S. EMPLOYEE TURNOVER RATE: In the U.S. call center, 11% of call center employees are expected to leave and be replaced in 2006. This turnover rate is expected to increase in future years.
- TRAINING COST PER U.S. EMPLOYEE: In the U.S. call center, it is expected to cost $1,100 to train a new employee in 2006, with the rate increasing in future years.
- TRAINING COST PER INDIAN EMPLOYEE: In the Indian call center, it is expected to cost $1,600 to train a new employee in 2006, with the rate increasing in future years.
- LEGAL AND TECHNICAL CONSULTING—INDIA: In 2005, BigState Bank expects to spend $15 million in legal fees, technical consulting fees, and PR to set up the off-shore operation. They expect to spend $6 million on maintaining the relationship in 2006, with the amount increasing each year in the future.
- TOTAL NUMBER OF LIAISON EMPLOYEES: Four liaison employees are expected in each year, two in each country.
- COST OF LIAISON EMPLOYEES: The cost of a liaison employee is expected to be $200,000 in 2006, increasing each year. This expense covers the cost of a person's salary, housing for the person's family, and so forth.
- COST OF EMPLOYEE BUYOUT: BigState Bank employees in the U.S. call center will be laid off at the end of 2005. To ensure loyalty in the last year of employment and to promote long-term goodwill in the community, a multi-year "buyout" amount will be paid. The amount will be $2,000 in 2006, decreasing each year, to zero in 2010.
- COST OF BACKGROUND CHECK—INDIA: A background check is expected to cost $1,000 in India per employee in 2006, increasing each year thereafter.
- COST OF BACKGROUND CHECK—U.S.: A background check is expected to cost $20 in the U.S. per employee in 2006, increasing each year thereafter.

INPUTS Section

Your spreadsheet should have the inputs shown in Figure 7-3. An explanation of the line items follows the figure.

	A	B	C	D	E	F	G
20	INPUTS	2005	2006	2007	2008	2009	2010
21	INDIA LABOR MARKET RISK: F = FAVORABLE, N = NOT FAVORABLE (ALL YEARS)		NA	NA	NA	NA	NA
22	EXCHANGE RATE: U = UP, S = STEADY, D = DOWN	NA					
23	DISCOUNT RATE FOR NPV		NA	NA	NA	NA	NA

Figure 7-3 INPUTS section

- INDIA LABOR MARKET RISK: The off-shoring agreement results are subject to labor market risks in India. Salary levels for information systems workers may go up a little, or a lot. Job turnover may go up a little, or a lot. Spreadsheet users who think these rates will not go up greatly in the years of the agreement would enter *F* for a "favorable" outcome. Pessimistic users will enter *N* for "not favorable." The entry applies to all the agreement's years.
- EXCHANGE RATE: If a user expects the dollar-to-rupee contract rate of 45.85 to remain relatively steady in the international money market during 2006–2010, the user enters an *S* for "Steady." If the user expects $1 to buy more than 45.85 rupees in 2006–2010, the user enters a *U* for "Up." If the user expects $1 to buy less than 45.85 rupees, the user enters *D* for "Down." An entry is made for each year of the agreement. The pattern of *SSSSS* would mean "Steady" in all years. The pattern *SSDDD* would mean "Steady at first, but then Down for three years."
- DISCOUNT RATE FOR NPV: The company's traditional hurdle rate is 25%, but this can be varied for a "what-if" analysis by entering another rate here. The entry applies to all the agreement's years.

Your instructor may tell you to apply Conditional Formatting to the input cells, so out-of-bounds values are highlighted in some way. (For example, the entry shows up in red type or in boldface type.) If so, your instructor may provide a tutorial on Conditional Formatting or may ask you to refer to Excel Help.

SUMMARY OF KEY RESULTS Section

Your spreadsheet should show one result, as shown in Figure 7-4.

	A	B
25	SUMMARY OF KEY RESULTS	2005-2010
26	NPV OF OFF-SHORING INVESTMENT	

Figure 7-4 SUMMARY OF KEY RESULTS section

The NPV of the off-shoring investment opportunity should be echoed to the Summary of Key Results section. The NPV value will vary, depending on inputs. There is one value covering the entire investment for the years 2005 to 2010. The key result cell should be formatted for currency, with no decimals.

CALCULATIONS Section

You should calculate various intermediate results, which are then used in other calculations, or in the net cash flow analysis that follows. Calculations are based on inputs or on year 2005 values. When called for, use absolute addressing properly. Calculate the values shown in Figures 7-5 and 7-6. An explanation of the line items follows the figures.

	A	B	C	D	E	F	G
28	CALCULATIONS	2005	2006	2007	2008	2009	2010
29	NUMBER OF U.S. EMPLOYEES	800					
30	NUMBER OF INDIAN EMPLOYEES	800					
31	NEW U.S. EMPLOYEES	NA					
32	INDIAN EMPLOYMENT TURNOVER RATE	NA					
33	NEW INDIAN EMPLOYEES	NA					
34	AVERAGE INDIAN EMPLOYEE EMPLOYMENT COST	16000					
35	NUMBER OF RUPEES PER U.S. DOLLAR	45.850					
36	INDIAN EMPLOYMENT COST:	NA	--	--	--	--	--
37	PAYMENT TO INDIAN CONTRACTOR (STEADY $$)	NA					
38	PAYMENT TO INDIAN CONTRACTOR (IN RUPEES)	NA					
39	DOLLARS NEEDED TO PAY CONTRACTOR IN RUPEES	NA					
40	COST OF LIAISON EMPLOYEES -- U.S.	NA					
41	INDIAN LIAISON EMPLOYEE COST:	NA	--	--	--	--	--
42	COST OF LIAISON EMPLOYEES -- INDIA	NA					
43	PAYMENT FOR INDIAN LIAISON EMPLOYEES (IN RUPEES)	NA					
44	DOLLARS NEEDED FOR INDIAN LIAISON PAYMENTS	NA					

Figure 7-5 CALCULATIONS section

- NUMBER OF U.S. EMPLOYEES: If the off-shoring agreement were not undertaken, this is the number of non-supervisory U.S. workers needed. There were 800 U.S. non-supervisory workers in 2005. That number would increase 2% per year due to expected banking business growth. Thus, the number of workers in 2006 would be 2% more than the number in 2005; the number in 2007 would be 2% more than in 2006, and so on. *Use the INT() function properly so that fractions of workers are not computed.*
- NUMBER OF INDIAN EMPLOYEES: If the off-shoring agreement is undertaken, this is the number of non-supervisory Indian workers needed. Assume that there would have been 800 in 2005 (1-to-1 replacement of U.S. workers). India's college graduates are expected to be more productive than U.S. call center workers, and the rate of increase is expected to be less. The number would increase 1% per year due to expected banking business growth. Thus, the number of workers in 2006 would be 1% more than the number in 2005; the number in 2007 would be 1% more than in 2006; and so on. *Use the INT() function properly so that fractions of workers are not computed.*
- NEW U.S. EMPLOYEES: New workers are hired for two reasons: (1) to accommodate business growth, and (2) to accommodate turnover (those who leave must be replaced). You can assume that turnover in a year is based on the number of workers in the prior year. For example: In Year 1, there were 900 workers. If the turnover rate is 10%, then the number of replacements to hire in Year 2 is 90.
- INDIAN EMPLOYMENT TURNOVER RATE: If the India labor market risk is expected to be Favorable, turnover will be only 10% per year during the life of the contract. If the risk is expected to be Not Favorable, the rate will be 25% per year.
- NEW INDIAN EMPLOYEES: The logic is the same as for new U.S. workers, except that in 2006 all Indian workers are new workers.
- AVERAGE INDIAN EMPLOYEE EMPLOYMENT COST: If the Indian labor market risk is expected to be Favorable, salary levels will increase only 8% a year during the life of the contract. If the risk is expected to be Not Favorable, the increase will be 15% a year.

- NUMBER OF RUPEES PER U.S. DOLLAR: If the exchange rate outlook is Steady, the number of rupees is the same as in the prior year. If the outlook is Up, the number of rupees per dollar will be 2% more than in the prior year. If the outlook is Down, the number will be 2% less than in the prior year.
- PAYMENT TO INDIAN CONTRACTOR (STEADY $$): This is based on the number of Indian workers and the average Indian worker employment cost. The amount is expressed in dollars.
- PAYMENT TO INDIAN CONTRACTOR (RUPEES): BI wants to be paid in rupees. This is the Payment to Indian Contractor, in rupees, at the agreed upon exchange rate of 45.85 in all years.
- DOLLARS NEEDED TO PAY CONTRACTOR IN RUPEES: This is the number of dollars actually required to pay for the Indian workers. This is based on the number of rupees needed (a calculation) and the number of rupees per U.S. dollar (another calculation).
- COST OF LIAISON EMPLOYEES—U.S.: This is based on the number of liaison employees living in the U.S. and the cost of a liaison employee. BigState Bank will pay this amount, in dollars.
- COST OF LIAISON EMPLOYEES—INDIA: This is based on the number of liaison employees living in India and the cost of a liaison employee.
- PAYMENT FOR INDIAN LIAISON EMPLOYEE (IN RUPEES): The liaison worker living in India will be paid by BigState Bank. The payment will be in rupees, calculated at the agreed 45.85 exchange rate in all years. The amount is in rupees.
- DOLLARS NEEDED FOR INDIAN LIAISON PAYMENTS: This is the number of dollars actually required to pay for the Indian liaison workers. This is based on the number of rupees needed (a calculation) and the number of rupees per U.S. dollar (another calculation).

	A	B	C	D	E	F	G
45	INDIAN BACKGROUND CHECK COST:	NA	--	--	--	--	--
46	COST OF BACKGROUND CHECKS -- INDIA	NA					
47	PAYMENT OF INDIAN BACKGROUND CHECKS (IN RUPEES)	NA					
48	DOLLARS NEEDED FOR INDIAN BACKGROUND CHECKS	NA					
49	INDIAN TRAINING COST:	NA	--	--	--	--	--
50	COST OF TRAINING INDIAN WORKERS	NA					
51	PAYMENT OF INDIAN TRAINING (RUPEES)	NA					
52	DOLLARS NEEDED FOR INDIAN TRAINING	NA					
53	COST OF BACKGROUND CHECKS -- U.S.	NA					
54	U.S. TELECOMM COSTS	NA					
55	U.S. COMPUTING COSTS	NA					
56	U.S. REAL ESTATE COSTS	NA					
57	U.S. TRAINING COSTS	NA					
58	NUMBER OF U.S. SUPERVISORS	NA					
59	U.S. SUPERVISORY COST	NA					
60	COST OF U.S. EMPLOYEES	NA					
61	COST OF EMPLOYEE BUYOUT	NA					

Figure 7-6 CALCULATIONS section (continued)

Case 7

- COST OF BACKGROUND CHECKS—INDIA: This is the cost of background checks, performed by BI, for new Indian workers in a year. The amount is expressed in dollars.
- PAYMENT OF INDIAN BACKGROUND CHECKS (IN RUPEES): BI wants to be paid in rupees, at the contract rate of 45.85 rupees per U.S. dollar, in all years.

- DOLLARS NEEDED FOR INDIAN BACKGROUND CHECKS: This is the number of dollars actually required to pay BI for the Indian background checks. This is based on the number of rupees needed (a calculation) and the number of rupees per U.S. dollar (another calculation)
- COST OF TRAINING INDIAN WORKERS: Each new Indian worker must be trained by BI, per the contract. The number of new workers is a calculation. The cost to train a worker (in dollars) is a constant.
- PAYMENT FOR INDIAN TRAINING (RUPEES): Employee training is for each new worker trained. BI wants to be paid in rupees. This is the cost of training new workers, at the contract rate of 45.85 per U.S. dollar, in all years.
- DOLLARS NEEDED FOR INDIAN TRAINING: This is the number of dollars actually required to pay for the Indian worker training. This is based on the number of rupees needed (a calculation) and the number of rupees per U.S. dollar (another calculation).
- COST OF BACKGROUND CHECKS—U.S.: This is the cost of background checks for new U.S. workers in a year.
- U.S. TELECOMM COSTS: This is based on the number of U.S. workers and the U.S. telecomm cost per worker.
- U.S. COMPUTING COSTS: This is based on the number of U.S. workers and the U.S. computer cost per worker.
- U.S. REAL ESTATE COSTS: This is based on the number of U.S. workers and the U.S. real estate cost per worker.
- U.S. TRAINING COSTS: This is based on the number of new U.S. workers and the cost to train a U.S. worker.
- NUMBER OF U.S. SUPERVISORS: This is based on the number of U.S. employees and the ratio of employees to supervisors.
- U.S. SUPERVISORY COST: This is based on the number of supervisors and the average supervisory employment cost.
- COST OF U.S. EMPLOYEES: This is based on the number of U.S. employees and the average employment cost for a U.S. worker.
- COST OF EMPLOYEE BUYOUT: This is based on the number of employees at the end of 2005 (who would be laid off) and the buyout amount in the year.

NET CASH FLOW ANALYSIS Section

The NPV calculation is based on net cash flows in the project's years. The net cash flow in a year is the difference between project benefits (cash inflows) less costs in the years (cash outflows). Values in this section should be formatted for no decimals. The section should look like Figure 7-7. Line items are discussed after the figure.

	A	B	C	D	E	F	G
63	**NET CASH FLOW ANALYSIS**						
64	**BENEFITS (CASH IN):**	**2005**	**2006**	**2007**	**2008**	**2009**	**2010**
65	FOREGONE U.S. COSTS	--	--	--	--	--	--
66	TELECOMM	**NA**					
67	COMPUTING	**NA**					
68	REAL ESTATE	**NA**					
69	TRAINING	**NA**					
70	SUPERVISORY SALARY	**NA**					
71	EMPLOYEE SALARY	**NA**					
72	BACKGROUND CHECKS	**NA**					
73	TOTAL BENEFITS	0					
74							
75	**COSTS (CASH OUT):**	**2005**	**2006**	**2007**	**2008**	**2009**	**2010**
76	DOLLARS NEEDED FOR PAYMENTS IN INDIA	--	--	--	--	--	--
77	CONTRACT FOR EMPLOYEES	**NA**					
78	LIAISON -- INDIA	**NA**					
79	LIAISON -- U.S.	**NA**					
80	BACKGROUND CHECKS	**NA**					
81	TRAINING	**NA**					
82	LEGAL AND TECHNICAL CONSULTING						
83	TOTAL COSTS						
84							
85	DIFFERENCE (BENEFITS - COSTS)						
86							
87	NPV OF OFF-SHORING INVESTMENT		**NA**	**NA**	**NA**	**NA**	**NA**

Figure 7-7 NET CASH FLOW ANALYSIS section

- BENEFITS (CASH IN): These values are amounts that are saved if the agreement is undertaken. The amounts are all calculations and can be echoed here. As you can see, benefits are totaled for each year. There are no benefits in 2005, so you can hard code a zero for that year's total benefits.
- COSTS (CASH OUT): These values are amounts that will be spent by BigState Bank if the agreement is undertaken. Most of the amounts are calculations and can be echoed here. Legal and Technical Consulting is a constant that can be echoed here. As you can see, costs are totaled for each year.
- DIFFERENCE (BENEFITS – COSTS): This is the difference in each year between total benefits and total costs in 2005 through 2010.
- NPV OF OFF-SHORING INVESTMENT: The NPV cell has the calculation of the project's NPV. *You should use the Excel Insert Function icon to research the built-in NPV() function.* Enter NPV in the input box, then follow the link to Excel Help to see explanations and examples. The NPV() function computes the NPV of an investment based on a discount rate and a series of one or more outflows (negative numbers) and a series of one or more inflows (positive numbers). The syntax is as follows:

  =NPV(discount rate, cell range for project net cash flows)

In this case, (1) the discount rate is in an input cell and (2) the net cash flow cells are in the Difference row in this section.

Case 7

ASSIGNMENT 2 USING THE SPREADSHEET FOR DECISION SUPPORT

You will now complete the case by (1) using the spreadsheet to gather the data needed to decide whether the agreement should be undertaken, (2) documenting your recommendation in a memorandum, and (3) making an oral presentation, if your instructor assigns it.

Assignment 2A: Using the Spreadsheet to Gather Data

You have built the spreadsheet to model the off-shoring decision. BigState Bank management wants to know what the NPV would be at a 25% hurdle rate in various scenarios. The scenarios are shown in the following shorthand notations:

- Favorable labor market outlook, exchange rate Up (Fav-Up)
- Favorable labor market outlook, exchange rate Steady (Fav-Steady)
- Favorable labor market outlook, exchange rate Down (Fav-Down)
- Not Favorable labor market outlook, exchange rate Up (NotFav-Up)
- Not Favorable labor market outlook, exchange rate Steady (NotFav-Steady)
- Not Favorable labor market outlook, exchange rate Down (NotFav-Down)

If NPVs are positive in all or most of the scenarios, management will probably go forward with the proposed agreement. However, if NPVs are negative in most or all of the scenarios, management will back out of the agreement—or seek modifications to its terms.

Now run "what-if" scenarios with the six sets of input values. The way you do this depends on whether your instructor has told you to use the Scenario Manager.

- If your instructor has told you *not* to use the Scenario Manager, you must now manually enter the input value combinations. Note the NPV result for each on a piece of paper (the value will show in the Summary of Key Results section as you work).
- If your instructor has told you to use the Scenario Manager, perform the procedures set forth in Tutorial C to set up and run the Scenario Manager. Record the six possible scenarios. The changing cells are the cells used to input the business risk outlook, exchange rate outlook, and the discount rate. In the Scenario Manager, you can enter non-contiguous cell ranges as follows: C20..F20, C21, C22 (cell addresses are arbitrary). The Output cell is the NPV value cell.
- In either case, when you are done gathering data, print the entire workbook (including the Scenario Summary sheet, if applicable). Then, save the spreadsheet (File—Save).

Assignment 2B: Documenting Your Recommendation in a Memorandum

Open MS Word, and write a brief memorandum to BigState Bank's CIO, who wants your view on what to do about the off-shoring opportunity. The goal is to have a positive NPV in all or most of the scenarios. The CIO would be reluctant to commit the bank to the agreement if NPVs were not positive. Here is guidance for your memorandum:

- Your memorandum should have a proper heading (DATE / TO / FROM / SUBJECT). You may wish to use a Word memo template (**File**, click **New**, click **On my computer** in the Templates section, click the **Memos** tab, chose **Contemporary Memo**, and click **OK**.).

- In the first paragraph, tell the CIO which scenarios result in a positive NPV and which do not, and state your recommendation about undertaking the agreement.
- Support your recommendation graphically, as your instructor requires: (1) If you used the Scenario Manager, go back into Excel, and put a copy of the Scenario Summary sheet results into the Windows Clipboard. Then, in Word, copy this graphic into the memorandum (Tutorial C refers to this procedure). (2) Make a summary table in Word after the first paragraph. The procedure is described next:

Enter a table into Word, using the following procedure:

1. Select the **Table** menu option, point to **Insert**, then click **Table**.
2. Enter the number of rows and columns.
3. Select **AutoFormat** and choose **Table Grid 1**
4. Select **OK**, and then select **OK** again.

Your table should resemble the format of the table shown in Figure 7-8:

Wage Level Outlook	Exchange Rate Outlook	Discount Rate	NPV
Favorable	Up	25%	
Favorable	Steady	25%	
Favorable	Down	25%	
Not Favorable	Up	25%	
Not Favorable	Steady	25%	
Not Favorable	Down	25%	

Figure 7-8 Format of table to insert in memorandum

ASSIGNMENT 3 GIVING AN ORAL PRESENTATION

Your instructor may request that you also present your analysis and recommendations in an oral presentation. If so, assume that the CIO has accepted your recommendation. He has asked you to give a presentation explaining your recommendation to the bank's senior management. Prepare to explain your analysis and recommendation to the group in 10 minutes or fewer. Use visual aids or handouts that you think are appropriate. Tutorial E has guidance on how to prepare and give an oral presentation.

DELIVERABLES

Assemble the following deliverables for your instructor:

1. Printout of your memorandum
2. Spreadsheet printouts
3. Disk or CD, which should have your Word memo file and your Excel spreadsheet file

Staple the printouts together, with the memorandum on top. If there is more than one .xls file on your disk or CD, write your instructor a note, stating the name of your model's .xls file.

8
CASE

The Starr Motor Company Survival Analysis

Decision Support Using Excel

Preview

Starr Motor Company is one of the many large companies in today's highly competitive automobile market. Starr has, unfortunately, been losing ground to competitors. In recent years, the company has improved its manufacturing efficiency and cut costs, and the company plans to introduce new auto and truck models in the next few years to stay competitive. However, it is unclear whether these moves will be sufficient for the company to remain competitive. In this case, you will use Excel to see whether the company can achieve profits and manage its high debt load.

Preparation

- Review spreadsheet concepts discussed in class and/or in your textbook. This case requires an understanding of the IF() and SUM() functions.
- Complete any exercises that your instructor assigns.
- Complete any part of Tutorial C that your instructor assigns, or refer to it as necessary.
- Review file-saving procedures for Windows programs. These are discussed in Tutorial C.
- Refer to Tutorial E as necessary.

BACKGROUND

Starr Motor Company has been making autos and trucks for more than 100 years. At one time, Starr was one of the dominant companies in the domestic automobile industry, making money hand over fist, year after year. Recent years, however, have been very difficult for Starr; indeed, they have been very difficult years for all automobile companies.

The automobile industry is now a global business. Twenty years ago, Starr competed against only two other auto companies in the domestic market, but there are now a dozen large companies, most headquartered in foreign lands. Each of these companies has good products, and many automobile industry executives now think that there is an oversupply of autos and trucks relative to consumer demand.

As any student of economics knows, when supply exceeds demand, there will be downward pressure on prices. Increased competition has other effects as well. These effects are discussed next.

Effect—Prices

In recent years, Starr's average unit selling price has been flat, even declining in some years. The primary mechanism for this has been so-called "incentive" programs. There are two kinds of incentives: cash-back incentives and special financing incentives. For example, Starr may list a car at $26,000, but in order to sell it, they must offer the potential buyer thousands of dollars in price discounts, or they may offer low-rate financing, or they may offer some combination of price discount and low-rate financing.

In recent years, most domestic auto companies have offered some level of incentives. Generally, this has meant dollars off the sticker price or low-rate financing. When times are especially competitive, the incentives can be almost ridiculous—companies might offer discounts of $5,000 on some models, or *zero* percentage-point financing. To exacerbate this problem, when Starr offers $5,000 off a $26,000 car, the $21,000 price is merely where the bargaining *begins* for the dealership selling the vehicle. Last year, Starr sold about 2.1 million vehicles at an average sales price under $19,000, which is surprisingly low when you consider the list price of a minivan, a nicely loaded family sedan, or a small truck.

Effect—Cost of Manufacture

There is a great deal of information in the popular press about vehicle reliability. Today's buyer will not consider buying an unreliable car or truck, so all manufacturers must now be concerned about quality. In the last decade, Starr has worked mightily to improve manufacturing quality and at the same time to lower the cost of making a vehicle. They have closed old manufacturing sites, and they have bargained hard with their unions on work rules, wages, and benefits. This effort has paid off: Quality is up, as evidenced by reduced warranty and recall costs; manufacturing costs are down—last year the average variable cost (labor and materials) to make a Starr vehicle was $15,000, which was lower than in the previous year. Management thinks that unit manufacturing costs will decrease about 1% per year in each of the next three years.

Effect—Research and Development

As always, auto companies are selling the excitement of a new car. Auto makers know that car models are popular for only so long, and then they must be replaced by better, and better looking, models. In recent years, Starr has been hard at work developing new cars. In the next three years, Starr plans to retire some old models and introduce new ones in all lines—sports cars, sedans, luxury cars, minivans, and trucks. (One new car is the limited edition RoadEater, a sports car that goes from zero to 60 in an amazing 2.9 seconds. As a marketing trick, Starr salespeople will be trained to administer a neck-strength physical to all potential buyers, to be sure they can handle the car's neck-snapping acceleration. Of course, the test is expected to show that all buyers who can handle the car's expected $100,000 list price have a sufficiently strong neck!) All other major car makers have been pumping money into new models too, but at least Starr management knows they will not be at an R&D disadvantage.

Better quality, lower costs, new models—that is the good news. The bad news for Starr has been that profit margins have not been good in recent years. The 2.1 million units sold last year was 7% less than the previous year. In fact, unit sales have been gradually creeping down for a number of years. This makes it increasingly difficult to cover fixed costs, marketing costs, general costs, and R&D expenses. Starr has very high debts. The company owes $20 billion to start the new year, and interest expense is high. In the last two years, Starr has about broken even—net income after taxes has been close to zero. Management knows that there is a great need to sell more cars.

Additional Sources of Revenue and Threats

In fact, Starr's most profitable activity in recent years has been *lending money* to car buyers. Like other companies, Starr has its own Finance Unit. This division borrows money at low rates in credit markets or from the company's bankers. Thus, when a person wants to buy a car and needs financing, the Finance Unit stands ready to lend money, usually at below-bank rates. A 48-month loan is created, collateralized by the vehicle. Recently, the Finance Unit has been able to borrow on average at 5%, and it has made loans to car buyers at 6% to 8%. Thus, the finance unit makes money for Starr in the form of interest earned on car loans. Of course, in periods when low-rate financing must be offered to buyers, the Finance Unit will not make money. In fact, it may borrow money at 5%, but then it must lend money at less than that rate. The incentives can really hurt the bottom line! The Finance Unit handles general corporate borrowing as well. For example, Starr may want to borrow $100 million to build a new manufacturing plant. Then, the Finance Unit would work with the company's bankers to borrow the money, or if the amount needed was huge, the Finance Unit would go to Wall Street to sell bonds.

Starr faces a threat from the financial markets that is related to its Financing Unit. All bond issuers are rated by independent credit analysis agencies. The ratings are intended as a measure of how likely a company is to pay off existing debts (interest and principle). If a company is very healthy, it will get a good rating. This increases investor confidence in the company and means that the company can borrow at lower rates. If a company is not very healthy, it will get a poor rating. A poor rating would reduce investor confidence and would mean that the company would have to borrow at high rates. Credit agency ratings can affect interest rates in the bond market and when borrowing directly from bankers. Currently, Starr has a poor credit rating, a B rating, but that rating is higher than the lowest bond rating, which is that of a "Junk bond." If the business worsens, however, the credit rating agencies probably would lower the rating to "Junk." Starr must borrow money every year to finance ongoing customer car sales, and Junk bond status would mean much higher interest expense for Starr.

There is other threatening news looming on the horizon, four or five years in the future. The Chinese have a state-subsidized automobile company that makes serviceable, but not too stylish, autos and trucks. Currently, they sell their products only in China, but that will soon change. The automaker is expanding capacity and revamping the styling, so they can compete directly in foreign markets with the world's dozen large auto makers. Chinese labor costs are lower than domestic labor costs. Competition from the Chinese vehicles is expected to drive down list prices for all auto makers.

Starr management expects to confront the Chinese auto company in four years. Management thinks the next three years are critical. This is management's reasoning: If Starr cannot turn things around now, and make large profits before the Chinese models are sold domestically, then perhaps management should dissolve the company. The company's better models could be sold to other makers and the proceeds distributed to the shareholders. This would be better than a slow march to bankruptcy, resulting in shareholders getting nothing.

You have been called in to make a spreadsheet model of Starr's financial prospects so management can decide what to do next.

Assignment 1 Creating a Spreadsheet for Decision Support

In this assignment, you will produce a spreadsheet that models the Starr automobile business. Then, in Assignment 2, you will write a memorandum to management about your analysis and your recommended action. In addition, in Assignment 3, you will be asked to prepare an oral presentation of your analysis and recommendation.

Next, you will create the spreadsheet model of Starr's business in the next three years. You will be given some hints on how each section should be set up before entering cell formulas. Your spreadsheet should have the following sections:

- **CONSTANTS**
- **INPUTS**
- **SUMMARY OF KEY RESULTS**
- **CALCULATIONS**
- **INCOME STATEMENT AND CASH FLOW STATEMENT**
- **DEBT OWED**

A discussion of each section follows. *The spreadsheet skeleton is available to you, so you need not type in the skeleton if you do not want to do so.* To access the spreadsheet skeleton, go to your Data files. Select Case 8, then select **AUTO.xls**.

CONSTANTS Section

Your spreadsheet should have the constants shown in Figure 8-1. An explanation of the line items follows the figure.

	A	B	C	D	E
1	**STARR MOTOR COMPANY**				
2					
3	**CONSTANTS**	**2005**	**2006**	**2007**	**2008**
4	TAX RATE	**NA**	0.2	0.2	0.2
5	MINIMUM CASH NEEDED TO START YEAR	**NA**	10,000,000	10,000,000	10,000,000
6	UNIT COST REDUCTION FACTOR	**NA**	0.01	0.01	0.01
7	FIXED COSTS		7,000,000,000	7,000,000,000	7,000,000,000

Figure 8-1 CONSTANTS section

- TAX RATE: The corporate tax rate is expected to be steady in the next three years.
- MINIMUM CASH NEEDED TO START YEAR: The company must have $10 million in cash on hand to start each year.
- UNIT COST REDUCTION FACTOR: Each car requires raw materials and direct labor during assembly. These are called "direct costs." The average value of a unit's direct costs is expected to go down 1% a year in each of the next three years. Thus, 2006's average direct cost per unit will be 1% less than 2005's, 2007's will be 1% less than 2006's, and so on.
- FIXED COSTS: Selling costs, general costs, and research-and-development costs are expected to be steady in the next three years at $7 billion a year.

INPUTS Section

Your spreadsheet should have the inputs shown in Figure 8-2. "NA" in a cell means that the cell should not be used for input values. An explanation of the line items follows the figure.

	A	B	C	D	E
9	**INPUTS**	**2005**	**2006**	**2007**	**2008**
10	CREDIT AGENCY RATING (B = WEAK; J = JUNK)		**NA**	**NA**	**NA**
11	UNIT SALES INCREASE (.XX)	**NA**			
12	INCENTIVES (S = STABLE; U = UP)	**NA**			

Figure 8-2 INPUTS section

- CREDIT AGENCY RATING: The current rating is a *B* rating in the bond market, and the rating is not expected to improve in the foreseeable future. The rating may soon fall to Junk status (*J*). The entry here applies to all three years.
- UNIT SALES INCREASE: The user enters the decimal percentage change in unit vehicle sales in a year. For example, if unit sales are expected to go up 5%, *.05* would be entered for the year. If sales were expected to be flat, *0* would be entered. If sales were to expect to decline 7%, *–.08* would be entered.
- INCENTIVES: Incentives are assumed to be a permanent feature of automobile marketing. If incentives are expected to be at normal levels, Stable (S) would be entered. However, occasionally companies compete intensively with incentives, such as very high price discounts or very low percentage financing. If aggressive incentives are expected in the year, Up (U) would be entered.

Your instructor may tell you to apply Conditional Formatting to the input cells, so that out-of-bounds values are highlighted in some way. (For example, the entry might be shown in red and/or in boldface type.) If so, your instructor may provide a tutorial on Conditional Formatting. Or, your instructor may ask you to refer to Excel Help.

SUMMARY OF KEY RESULTS Section

Your spreadsheet should show one result, as shown in Figure 8-3.

	A	B	C	D	E
14	SUMMARY OF KEY RESULTS	2005	2006	2007	2008
15	NET INCOME AFTER TAXES	NA			
16	END-OF-THE-YEAR CASH ON HAND	NA			
17	END-OF-THE-YEAR DEBT OWED	NA			

Figure 8-3 SUMMARY OF KEY RESULTS section

For each year, your spreadsheet should show (1) net income after taxes for the year, (2) cash on hand at the end of the year, and (3) debt owed at the end of the year to bankers and bond holders. These values are all computed elsewhere in the spreadsheet and can be echoed here. These cells should be formatted for zero decimals.

CALCULATIONS Section

You should calculate various intermediate results, which are then used in the INCOME STATEMENT AND CASH FLOW STATEMENT section. Calculations, shown in Figure 8-4, are based on inputs and/or on year 2005 values. When called for, use absolute addressing. An explanation of the line items follows the figure.

	A	B	C	D	E
19	CALCULATIONS	2005	2006	2007	2008
20	INTEREST RATE ON DEBT	NA			
21	UNITS SOLD	2125000			
22	SELLING PRICE PER UNIT	18700			
23	COST TO MAKE A UNIT	15000			
24	INTEREST EARNED PER UNIT SOLD	NA			

Figure 8-4 CALCULATIONS section

- INTEREST RATE ON DEBT: If the credit agency rating will be *B* in the three years, the interest rate paid on debt owed will be 5% in each of the three years. If the rating moves down to Junk-bond status, however, the rate will be 10% in each of the three years.
- UNITS SOLD: The number of vehicles sold in a year is based on the prior year's units and the unit sales increase factor (an input value).
- SELLING PRICE PER UNIT: If the level of incentives is expected to be Stable (*S*) in a year, then Starr can be expected to raise the average unit selling price 1% over the prior year's price. However, if incentives are expected to be Up (*U*) in a year, the average selling price in a year will be 5% less than the prior year's selling price.
- COST TO MAKE A UNIT: The average direct cost to make a vehicle in a year is based on the prior year's direct cost to make a unit, and on the unit cost reduction factor (a constant).

- INTEREST EARNED PER UNIT SOLD: Not all car buyers will finance the purchase through Starr, but the majority of buyers will. Financing is a source of income to Starr. If incentives are stable in a year, the company can expect to make $150 in interest revenue on a unit sold, on average. However, if incentives are up, only $20 in interest revenue is earned per unit sold, on average.

Income Statement And Cash Flow Statement

The forecast for net income and cash flow starts with the cash on hand at the beginning of the year. This is followed by the income statement and concludes with the calculation of cash on hand at year-end. For readability, format cells in this section for zero decimals. Your spreadsheet should look like the ones shown in Figures 8-5 and 8-6. A discussion of the line items follows the figures.

	A	B	C	D	E
26	INCOME STATEMENT AND CASH FLOW STATEMENT	2005	2006	2007	2008
27	BEGINNING-OF-THE-YEAR CASH ON HAND				
28					
29	REVENUE		--	--	--
30	AUTO SALES	NA			
31	INTEREST EARNED	NA			
32	TOTAL REVENUE	NA			
33	COSTS AND EXPENSES		--	--	--
34	COST OF AUTOS SOLD	NA			
35	FIXED COSTS	NA			
36	TOTAL COSTS AND EXPENSES	NA			
37	INCOME BEFORE INTEREST AND TAXES	NA			
38	INTEREST EXPENSE	NA			
39	INCOME BEFORE TAXES	NA			
40	INCOME TAX EXPENSE	NA			
41	NET INCOME AFTER TAXES	NA			

Figure 8-5 INCOME STATEMENT AND CASH FLOW STATEMENT section

- BEGINNING-OF-THE-YEAR CASH ON HAND: This is the cash on hand at the end of the *prior* year.
- REVENUE—AUTO SALES: This is based on units sold and on the average selling price in the year. These amounts are calculations.
- REVENUE—INTEREST EARNED: This is based on units sold and on the interest earned per unit sold in a year. These amounts are calculations.
- TOTAL REVENUE: This is the total of auto sales and interest earned.
- COST OF AUTOS SOLD: This is based on units sold and on the average direct cost to make a unit. These amounts are calculations.
- FIXED COSTS: This amount is a constant that can be echoed here.
- TOTAL COSTS AND EXPENSES: This is the total of the direct cost of autos sold and of fixed costs.
- INCOME BEFORE INTEREST AND TAXES: This is the difference between total revenue and total costs and expenses.
- INTEREST EXPENSE: This is based on the year's interest rate (a calculation) and the amount owed at the beginning of the year.
- INCOME BEFORE TAXES: This is the difference between income before interest and taxes and interest expense.

- INCOME TAX EXPENSE: This is zero if income before taxes is zero or less; otherwise, apply the tax rate for the year to income before taxes to calculate income tax expense.
- NET INCOME AFTER TAXES: This is the difference between income before taxes and income tax expense.

Continuing this statement, line items for the year-end cash calculation are discussed. In Figure 8-6, column B is for 2005, column C for 2006, and so on.

	A	B	C	D	E
43	NET CASH POSITION (NCP) BEFORE BORROWING AND REPAYMENT OF DEBT (BEG OF YR CASH + NET INCOME)	NA			
44	ADD: BORROWING FROM BANK	NA			
45	LESS: REPAYMENT TO BANK	NA			
46	EQUALS: END-OF-THE-YEAR CASH ON HAND	10,000,000			

Figure 8-6 END-OF-THE-YEAR CASH ON HAND section

- Year 2005 values are mostly NA, except that END-OF-THE-YEAR CASH ON HAND for 2005 was $10 million, which is Starr's minimum cash requirement.
- The NET CASH POSITION (NCP) at the end of a year equals cash at the beginning of the year, plus the year's net income after taxes.
- ADD: BORROWING FROM BANK: Assume that Starr's bankers will lend the company enough money at year-end to get to the minimum cash needed to start the next year. If NCP is less than the minimum cash needed, Starr must borrow enough money to get to the minimum. Borrowings increase cash on hand, of course.
- LESS: REPAYMENT TO BANK: If the NCP is more than the minimum cash at the end of a year and there is debt owed, Starr must then pay off as much debt as possible (but not take the company below the minimum cash required to start the year). Here, "bank" refers to bankers and bond holders. Repayments reduce cash on hand, of course.
- END-OF-THE-YEAR CASH ON HAND: This equals the NCP plus any borrowings, less any repayments.

DEBT OWED Section

Your spreadsheet body ends with a calculation of debt owed at year-end, as shown in Figure 8-7. An explanation of the line items follows the figure.

	A	B	C	D	E
48	DEBT OWED	2005	2006	2007	2008
49	BEGINNING-OF-THE-YEAR DEBT OWED	NA			
50	ADD: BORROWING FROM BANK	NA			
51	LESS: REPAYMENT TO BANK	NA			
52	EQUALS: END-OF-THE-YEAR DEBT OWED	20,000,000,000			

Figure 8-7 DEBT OWED section

- Year 2005 values are mostly NA, except that the company owes $20 billion to bankers and bond holders at the end of 2005.
- BEGINNING-OF-THE-YEAR DEBT OWED: Cash owed at the beginning of a year equals cash owed at the end of the prior year.

- ADD: BORROWING FROM BANK, LESS: REPAYMENT TO BANK: These amounts have been calculated and can be echoed in this section.
- EQUALS: END-OF-THE-YEAR DEBT OWED: This equals the amount owed at the beginning of a year, plus borrowings in the year, less repayments in the year.

ASSIGNMENT 2 USING THE SPREADSHEET FOR DECISION SUPPORT

You will now complete the case by (1) using the spreadsheet to gather the data needed to assess Starr's financial prospects, and (2) documenting your recommendation in a memorandum.

Assignment 2A: Using the Spreadsheet to Gather Data

You have built the spreadsheet to model Starr's financial prospects in the next three years. Management wants to know financial results in four conceivable scenarios:

1. **Optimistic:** The credit rating remains weak (*B*), unit sales increase 2% a year (.02), and incentives remain Stable (*S*). This scenario would clearly represent an improvement in the company's affairs, inasmuch as unit sales have been declining in recent years.
2. **Stable:** The credit rating remains weak (*B*), unit sales remain level (*0*), and incentives remain Stable (*S*). This is an improvement over where the company is now—the unit sales bleeding has stopped.
3. **Trouble:** The credit rating remains weak (*B*), unit sales decrease 2% a year (.02), and incentives remain Stable (*S*). This is about where the company is now, including a decline in unit sales.
4. **Worst Case:** The credit rating goes down to Junk bond status (*J*), unit sales decrease 5% a year (.05), and incentives are forced to go up (*U*) for the company to remain competitive. This would probably be a very bad situation for the company.

The financial status in 2008 is the key, in management's view. For each scenario, Management wants to know the following the following:

1. The 2008 net income
2. Cash on hand at the end of 2008
3. Debt owed at the end of 2008

Management knows that the company cannot be debt-free in 2008, but it would like to see significant debt reductions in the next three years. This is because the company will probably need to take on more debt later to counter the threat from Chinese imports. Management will use net income to reduce debt in the next three years. A debt of $15 billion or less at the end of 2008 would probably be considered very good news in the financial markets. A debt level over $18 billion would probably be considered bad news, but not necessarily life threatening.

Management members wonder how different the results are from one scenario to the next. Put another way: as long as the worst-case scenario is avoided, is the company in about the same place in each of the other three scenarios? If that were the case, and the results were otherwise acceptable, then the strategy would be to do all things possible in order to avoid the worst-case scenario conditions (Junk status, great decrease in units sold, excessive incentives).

Now run "what-if" scenarios with the four sets of input values. The way you do this depends on whether your instructor has told you to use the Scenario Manager.

If your instructor has told you *not* to use the Scenario Manager, you must now manually enter the input value combinations. Note the results for each on a piece of paper (the value will show in the SUMMARY OF KEY RESULTS section as you work). Additionally, if your instructor has told you to chart any results, manually enter the data into a separate charting area, perhaps to the right of the SUMMARY OF KEY RESULTS section or below the DEBT OWED section.

If your instructor has told you to use the Scenario Manager, perform the procedures set forth in Tutorial C to set up and run the Scenario Manager. Record the four possible scenarios. The changing cells are the cells used to input the credit agency rating, the unit sales increase, and the incentives status. In the Scenario Manager, you can enter non-contiguous cell ranges as follows: C20..F20, C21, C22 (cell addresses are arbitrary). The Output cells are the 2008 cells in the Summary of Key Results. If your instructor has told you to chart any results, you can chart the Scenario Manager Summary sheet values, which are nicely arrayed for charting purposes.

In either case, when you are done gathering data, print the entire workbook (including the Scenario Summary sheet, if applicable). Then, save the spreadsheet (File—Save).

Assignment 2B: Documenting Your Recommendation in a Memorandum

Open MS Word, and write a brief memorandum to Starr's CEO. Assume that the CEO wants to know about your analysis and wants your interpretation of the results. Here are guidelines on writing your memorandum:

- Your memorandum should have a proper heading (DATE / TO / FROM / SUBJECT). You may wish to use a Word memo template (**File**, click **New**, click **On my computer** in the Templates section, click the **Memos** tab, choose **Contemporary Memo**, and click **OK**.)
- You need not provide background—the CEO is quite aware of the situation. You should summarize your analytical method and state the results. Then, you should give your interpretation of the results, relative to the debt situation. Discuss each scenario's results. Be sure to discuss how likely the company is to avoid the worst-case scenario.
- Support your statements graphically, as your instructor requires: (1) If you used the Scenario Manager, your instructor may want you to go back into Excel, and put a copy of the Scenario Manager Summary sheet results into the Windows Clipboard. Then, in Word, copy this graphic into the memorandum (Tutorial C refers to this procedure). (2) If your instructor had you make any charts, either insert them into the memorandum or attach them as separate exhibits, as your instructor specifies. (3) Or, your instructor may want you to make a summary table in Word, based on the Scenario Manager Summary sheet, after the first paragraph. The procedure for creating a table in Word is described next.

Enter a table into Word, using the following procedure:

1. Select the **Table** menu option, click **Insert**, and then click **Table**.
2. Enter the number of rows and columns.
3. Select AutoFormat and choose **Table Grid 1.**
4. Select **OK**, and then select **OK** again.

Your table in this case should resemble the format of the table shown in Figure 8-8.

Scenario	2008 Net Income After Taxes	2008 End-of-the-Year Cash	2008 End-of-the-Year Debt Owed
Optimistic			
Stable			
Trouble			
Worst Case			

Figure 8-8 Format of table to insert in memorandum

ASSIGNMENT 3 GIVING AN ORAL PRESENTATION

Your instructor may request that you also present your analysis and recommendations in an oral presentation. If so, assume that the CEO is impressed by your method and by your findings. He has asked you to give a presentation explaining your results to the company's management and to its creditors. Prepare to explain your analysis and recommendation to the group in 10 minutes or fewer. Use visual aids or handouts that you think are appropriate. Tutorial E has guidance on how to prepare and give an oral presentation.

DELIVERABLES

1. Printout of your memorandum
2. Spreadsheet printouts
3. Disk or CD, which should have your Word memo file and your Excel spreadsheet file

Staple the printouts together, with the memorandum on top. If there is more than one .xls file on your disk or CD, write your instructor a note, stating the name of your model's .xls file.

Decision Support Cases Using the Excel Solver

TUTORIAL

Building a Decision Support System Using the Excel Solver

Decision Support Systems (DSS) help people to make decisions. (The nature of DSS programs is discussed in Tutorial C.) Tutorial D teaches you how to use the Solver, one of Excel's built-in decision support tools.

For some business problems, decision makers want to know the best, or optimal, solution. Usually this means maximizing a variable (e.g., net income) or minimizing another variable (e.g., total costs). This optimization is subject to constraints, which are rules that must be observed when solving a problem. The Solver computes answers to such optimization problems.

This tutorial has four sections:

1. **Using the Excel Solver** In this section, you'll learn how to use the Solver in decision making. As an example, you use the Solver to create a production schedule for a sporting goods company. This schedule is called the Base Case.
2. **Extending the Example** In this section, you'll test what you've learned about using the Solver as you modify the sporting goods company's production schedule. This is called the Extension Case.
3. **Using the Solver on a New Problem** In this section, you'll use the Solver on a new problem.
4. **Trouble-shooting the Solver** In this section, you'll learn how to overcome problems you might encounter when using the Solver.

Tutorial C has some guidance on basic Excel concepts, such as formatting cells and using functions, such as =IF(). Refer to Tutorial C for a review of such topics.

Using the Excel Solver

Suppose that a company must set a production schedule for its various products, each of which has a different profit margin (selling price less costs). At first, you might assume that the company will maximize production of all profitable products to maximize net income. However, a company typically cannot make and sell an unlimited number of its products because of constraints.

One constraint affecting production is the "shared resource problem." For example, several products in a manufacturer's line might require the same raw materials, which are in limited supply. Similarly, the manufacturer might require the same machines to make several of its products. In addition, there might also be a limited pool of skilled workers available to make the products.

In addition to production constraints, sometimes management's policies impose constraints. For example, management might decide that the company must have a broader product line. As a consequence, a certain production quota for several products must be met, regardless of profit margins.

Thus, management must find a production schedule that will maximize profit, given the constraints. Optimization programs like the Solver look at each combination of products, one after the other, ranking each combination by profitability. Then the program reports the most profitable combination.

To use the Solver, you'll set up a model of the problem, including the factors that can vary, the constraints on how much they can vary, and the goal you are trying to maximize (usually net income) or minimize (usually total costs). The Solver then computes the best solution.

Setting Up a Spreadsheet Skeleton

Suppose that your company makes two sporting goods products—basketballs and footballs. Assume that you will sell all the balls you produce. To maximize net income, you want to know how many of each kind of ball to make in the coming year.

Making each kind of ball requires a certain (and different) number of hours, and each ball has a different raw materials cost. Because you have only a limited number of workers and machines, you can devote a maximum of 40,000 hours to production. This is a shared resource. You do not want that resource to be idle, however. Downtime should be no more than 1,000 hours in the year, so machines should be used for at least 39,000 hours.

Marketing executives say you cannot make more than 60,000 basketballs and may not make less than 30,000. Furthermore, you must make at least 20,000 footballs but not more than 40,000. Marketing says the ratio of basketballs to footballs produced should be between 1.5 and 1.7—i.e., more basketballs than footballs, but within limits.

What would be the best production plan? This problem has been set up in the Solver. The spreadsheet sections are discussed in the pages that follow.

AT THE KEYBOARD

Start out by saving the blank spreadsheet as **SPORTS1.xls**. Then you should enter the skeleton and formulas as they are discussed.

CHANGING CELLS Section

The **CHANGING CELLS** section contains the variables the Solver is allowed to change while it looks for the solution to the problem. Figure D-1 shows the skeleton of this spreadsheet section and the values that you should enter. An analysis of the line items follows the figure.

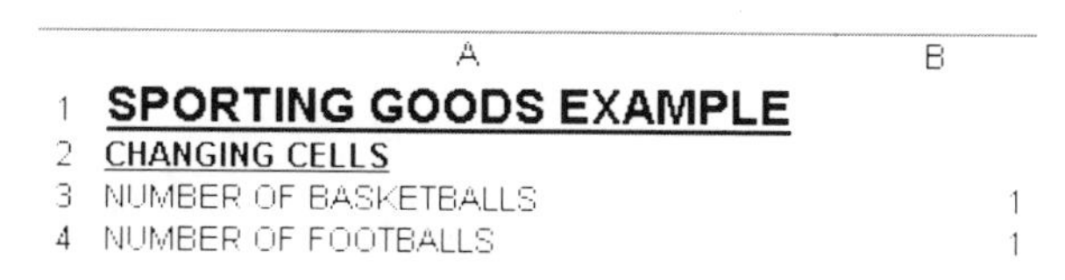

	A	B
1	SPORTING GOODS EXAMPLE	
2	CHANGING CELLS	
3	NUMBER OF BASKETBALLS	1
4	NUMBER OF FOOTBALLS	1

Figure D-1 CHANGING CELLS section

Tutorial D

- The changing cells are for the number of basketballs and footballs to be made and sold. The changing cells are like input cells, except Solver (not you) plays "what-if" with the values, trying to maximize or minimize some value (in this case, maximize net income).
- Note that some number should be put in the changing cells each time before the Solver is run. It's customary to put the number 1 into the changing cells (as shown). Solver will change these values when the program is run.

CONSTANTS Section

Your spreadsheet should also have a section for values that will not change. Figure D-2 shows a skeleton of the **CONSTANTS** section and the values you should enter. A discussion of the line items follows the figure.

You should use Format—Cells—Number to set the constants range to two decimal places.

	A	B
6	CONSTANTS	
7	BASKETBALL SELLING PRICE	14.00
8	FOOTBALL SELLING PRICE	11.00
9	TAX RATE	0.28
10	NUMBER OF HOURS TO MAKE A BASKETBALL	0.50
11	NUMBER OF HOURS TO MAKE A FOOTBALL	0.30
12	COST OF LABOR -- 1 MACHINE HOUR	10.00
13	COST OF MATERIALS -- 1 BASKETBALL	2.00
14	COST OF MATERIALS -- 1 FOOTBALL	1.25

Figure D-2 CONSTANTS section

- The SELLING PRICE for one basketball and for one football is shown.
- The TAX RATE is the rate applied to income before taxes to compute income tax expense.
- The NUMBER OF MACHINE HOURS needed to make a basketball and a football is shown. Note that a ball-making machine can produce two basketballs in an hour.
- COST OF LABOR: A ball is made by a worker using a ball-making machine. A worker is paid $10 for each hour he or she works at a machine.
- COST OF MATERIALS: The costs of raw materials for a basketball and football are shown.

Notice that the profit margins (selling price less costs of labor and materials) for the two products are not the same. They have different selling prices and different inputs (raw materials, hours to make) and the inputs have different costs per unit. Also note that you cannot tell from the data how many hours of the shared resource (machine hours) will be devoted to basketballs and how many to footballs, because you don't know in advance how many basketballs and footballs will be made.

CALCULATIONS Section

In the **CALCULATIONS** section, you will calculate intermediate results that (1) will be used in the spreadsheet body, and/or (2) will be used as constraints. First, use Format—Cells—Number to set the calculations range to two decimal places. Figure D-3 shows the skeleton and formulas that you should enter. A discussion of the cell formulas follows the figure.

Cell widths are changed here merely to show the formulas—you need not change the width.

	A	B
16	CALCULATIONS	
17	RATIO OF BASKETBALLS TO FOOTBALLS	=B3/B4
18	TOTAL BASKETBALL HOURS USED	=B3*B10
19	TOTAL FOOTBALL HOURS USED	=B4*B11
20	TOTAL MACHINE HOURS USED (BB + FB)	=B18+B19

Figure D-3 CALCULATIONS section cell formulas

- The RATIO OF BASKETBALLS TO FOOTBALLS (cell B17) will be needed in a constraint.
- TOTAL BASKETBALL HOURS USED: The number of machine hours needed to make all basketballs (B3 * B10) is computed in cell B18. Cell B10 has the constant for the hours needed to make one basketball. Cell B3 (a changing cell) has the number of basketballs made. (Currently, this cell shows one ball, but that number will change when the Solver works on the problem.)
- TOTAL FOOTBALL HOURS USED: The number of machine hours needed to make all footballs is calculated similarly, in cell B19.
- TOTAL MACHINE HOURS USED (BB + FB): The number of hours needed to make both kinds of balls (cell B20) will be a constraint; this value is the sum of the hours just calculated for footballs and basketballs.

Notice that constants in the Excel cell formulas in Figure D–3 are referred to by their cell addresses. Use the cell address of a constant rather than hard-coding a number in the Excel expression: If the number must be changed later, you only have to change it in the CONSTANTS section cell, not in every cell formula in which you used the value.

Notice that you do not calculate the amounts in the changing cells (here, the number of basketballs and footballs to produce). The Solver will compute those numbers. Also notice that you can use the changing cell addresses in your formulas. When you do that, you assume the Solver has put the optimal values in each changing cell; your expression makes use of that number.

Figure D-4 shows the values after Excel evaluates the cell formulas (with 1s in the changing cells):

	A	B
16	CALCULATIONS	
17	RATIO OF BASKETBALLS TO FOOTBALLS	1.00
18	TOTAL BASKETBALL HOURS USED	0.50
19	TOTAL FOOTBALL HOURS USED	0.30
20	TOTAL MACHINE HOURS USED (BB + FB)	0.80

Figure D-4 CALCULATIONS section cell values

INCOME STATEMENT Section

The target value is calculated in the spreadsheet body in the INCOME STATEMENT section. This is the value that the Solver is expected to maximize or minimize. The spreadsheet body can take any form. In this textbook's Solver cases, the spreadsheet body will be an income statement. Figure D-5 shows the skeleton and formulas that you should enter. A discussion of the line-item cell formulas follows the figure.

NOTE

Income statement cells were formatted for two decimal places.

	A	B
22	INCOME STATEMENT	
23	BASKETBALL REVENUE (SALES)	=B3*B7
24	FOOTBALL REVENUE (SALES)	=B4*B8
25	TOTAL REVENUE	=B23+B24
26	BASKETBALL MATERIALS COST	=B3*B13
27	FOOTBALL MATERIALS COST	=B4*B14
28	COST OF MACHINE LABOR	=B20*B12
29	TOTAL COST OF GOODS SOLD	=SUM(B26:B28)
30	INCOME BEFORE TAXES	=B25-B29
31	INCOME TAX EXPENSE	=IF(B30<=0,0,B30*B9)
32	NET INCOME AFTER TAXES	=B30-B31

Figure D-5 INCOME STATEMENT section cell formulas

- REVENUE (cells B23 and B24) equals the number of balls times the respective unit selling price. The number of balls is in the changing cells, and the selling prices are constants.
- MATERIALS COST (cells B26 and B27) follows a similar logic: number of units times unit cost.
- The COST OF MACHINE LABOR is the calculated number of machine hours times the hourly labor rate for machine workers.
- TOTAL COST OF GOODS SOLD is the sum of the cost of materials and the cost of labor.

- This is the logic of income tax expense: If INCOME BEFORE TAXES is less than or equal to zero, the tax is zero; otherwise, the income tax expense equals the tax rate times income before taxes. An =IF() statement is needed in cell B31.

Excel evaluates the formulas. Figure D-6 shows the results (assuming 1s in the changing cells):

	A	B
22	**INCOME STATEMENT**	
23	BASKETBALL REVENUE (SALES)	14.00
24	FOOTBALL REVENUE (SALES)	11.00
25	TOTAL REVENUE	25.00
26	BASKETBALL MATERIALS COST	2.00
27	FOOTBALL MATERIALS COST	1.25
28	COST OF MACHINE LABOR	8.00
29	TOTAL COST OF GOODS SOLD	11.25
30	INCOME BEFORE TAXES	13.75
31	INCOME TAX EXPENSE	3.85
32	NET INCOME AFTER TAXES	9.90

Figure D-6 INCOME STATEMENT section cell values

Constraints

Constraints are rules which the Solver must observe when computing the optimal answer to a problem. Constraints will need to refer to calculated values, or to values in the spreadsheet body. Therefore, you must build those calculations into the spreadsheet design, so they are available to your constraint expressions. (There is no section on the face of the spreadsheet for constraints. You'll use a separate window to enter constraints.)

Figure D-7 shows the English and Excel expressions for the basketball and football production problem constraints. A discussion of the constraints follows the figure.

Expression in English	Excel Expression
TOTAL MACHINE HOURS >= 39000	B20 >= 39000
TOTAL MACHINE HOURS <= 40000	B20 <= 40000
MIN BASKETBALLS = 30000	B3 >= 30000
MAX BASKETBALLS = 60000	B3 <= 60000
MIN FOOTBALLS = 20000	B4 >= 20000
MAX FOOTBALLS = 40000	B4 <= 40000
RATIO BB'S TO FB'S-MIN = 1.5	B17 >= 1.5
RATIO BB'S TO FB'S-MAX = 1.7	B17 <= 1.7
NET INCOME MUST BE POSITIVE	B32 >= 0

Figure D-7 Solver Constraint expressions

- As shown in Figure D-7, notice that a cell address in a constraint expression can be a cell address in the **CHANGING CELLS** section, a cell address in the **CONSTANTS** section, a cell address in the **CALCULATIONS** section, or a cell address in the spreadsheet body.
- You'll often need to set minimum and maximum boundaries for variables. For example, the number of basketballs (MIN and MAX) varies between 30,000 and 60,000 balls.
- Often, a boundary value is zero because you want the Solver to find a non-negative result. For example, here you want only answers that yield a positive net income. You tell the Solver that the amount in the net income cell must equal or exceed zero, so the Solver does not find an answer that produces a loss.
- Machine hours must be shared between the two kinds of balls. The constraints for the shared resource are: B20 >= 39000 and B20 <= 40000, where cell B20 shows the total hours used to make both the basketballs and footballs. The shared-resource constraint seems to be the most difficult kind of constraint for students to master when learning the Solver.

Running the Solver: Mechanics

To set up the Solver, you must tell the Solver these things:

1. The cell address of the "target" variable that you are trying to maximize (or minimize, as the case may be)
2. The changing cell addresses
3. The expressions for the constraints

The Solver will put its answers in the changing cells and on a separate sheet.

Beginning to Set Up the Solver

AT THE KEYBOARD

To start setting up the Solver, select Tools—Solver. The first thing you will see is a Solver Parameters window, as shown in Figure D-8. Use the Solver Parameters window to specify the target cell, the changing cells, and the constraints. If you don't see the Solver tool under the Tools menu, you may need to activate it by going to Tools—Add-Ins and clicking the Solver Add-Ins box to install it.

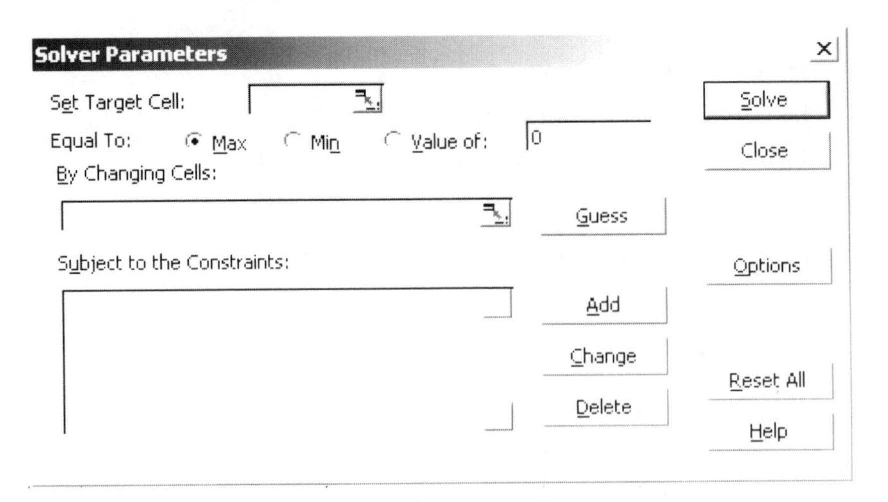

Figure D-8 Solver Parameters window

Setting the Target Cell

To set a target cell, use the following procedure:

1. The Target Cell is net income, cell B32.
2. To set the Target Cell, click in that input box and enter B32.
3. Max is the default; accept it here.
4. Enter a "0" for no desired net income value (Value of). DO NOT hit Enter when you finish. You'll navigate within this window by clicking in the next input box.

Figure D-9 shows entering data in the Target Cell.

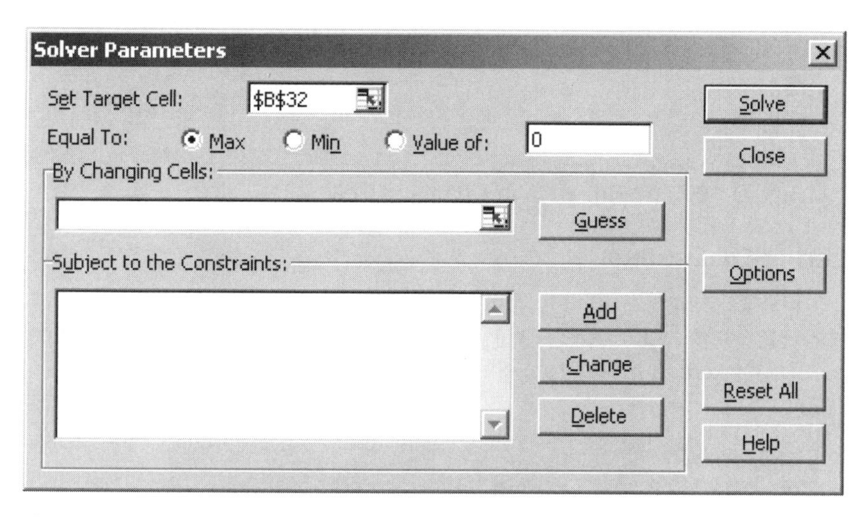

Figure D-9 Entering data in the Target Cell

When you enter the cell address, Solver may put in dollar signs, as if for absolute addressing. Ignore them—do not try to delete them.

Setting the Changing Cells

The changing cells are the cells for the balls, which are in the range of cells B3:B4. Click in the Changing Cells box and enter B3:B4, as shown in Figure D10. (Do *not* then hit Enter.)

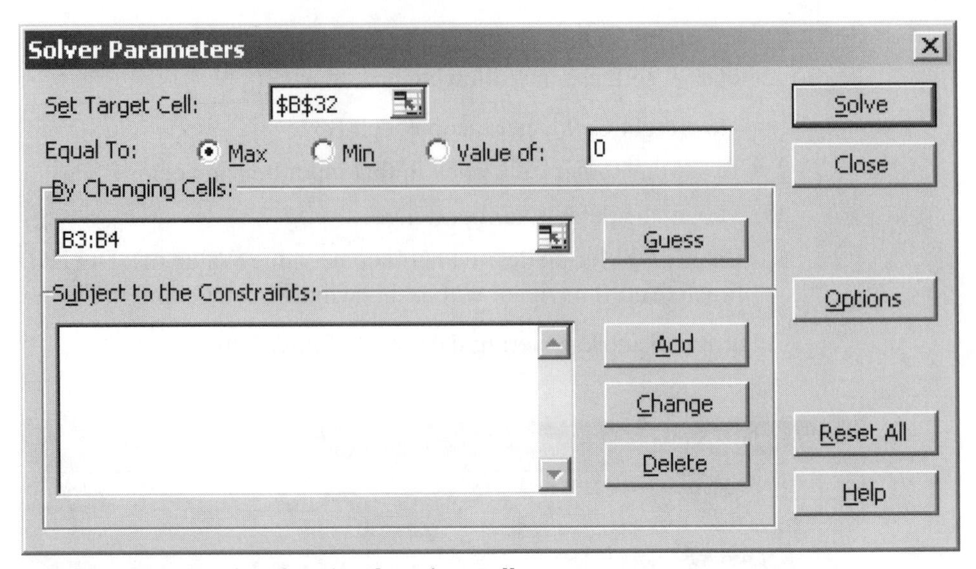

Figure D-10 Entering data in Changing Cells

Entering Constraints

You are now ready to enter the constraint formulas one by one. To start, click the Add button. As shown in Figure D-11, you'll see the Add Constraint window (here, shown with the minimum basketball production constraint entered).

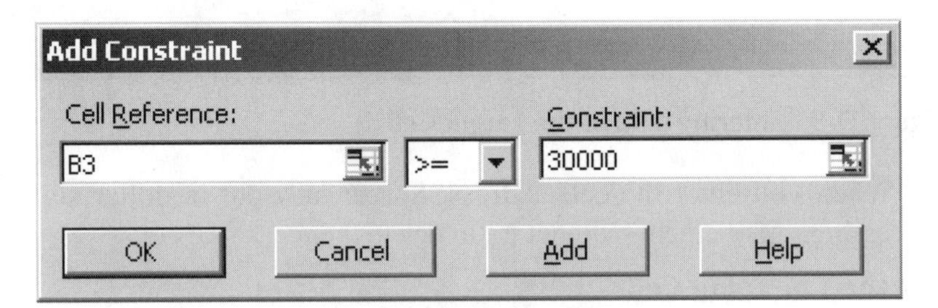

Figure D-11 Entering data in Add Constraint box

You should note the following about entering constraints and Figure D-11:

- To enter a constraint expression, do four things: (1) Type the variable's cell address in the left Cell Reference input box; (2) select the operator (<=, =, >=) in the smaller middle box; (3) enter the expression's right-hand side value, which is either a raw number or the cell address of a value, into the Constraint box; and (4) click Add to enter the constraint into the program. If you change your mind about the expression and do not want to enter it, click Cancel.
- The minimum basketballs constraint is: B3 >= 30000. Enter that constraint now. (Later, Solver may put an "equals" sign in front of the 30000 and dollar signs in the cell reference.)
- After entering the constraint formula, click the Add button. This puts the constraint into the Solver model. It also leaves you in the Add Constraint window, allowing you to enter other constraints. You should enter those now. See Figure D-7 for the logic.
- When you're done entering constraints, click the Cancel button. This takes you back to the Solver Parameters window.

You should not put an expression into the Cell Reference window. For example, the constraint for the minimum basketball-to-football ratio is B3/B4 >= 1.5. You should not put =B3/B4 into the Cell Reference box. This is why the ratio is computed in the Calculations section of the spreadsheet (in cell B17). When adding that constraint, enter B17 in the Cell Reference box. (You are allowed to put an expression into the Constraint box, although that technique is not shown here and is not recommended.)

After entering all the constraints, you'll be back at the Solver Parameters window. You will see the constraints have been entered into the program. Not all constraints will show, due to the size of the box. The top part of the box's constraints area looks like the portion of the spreadsheet shown in Figure D-12.

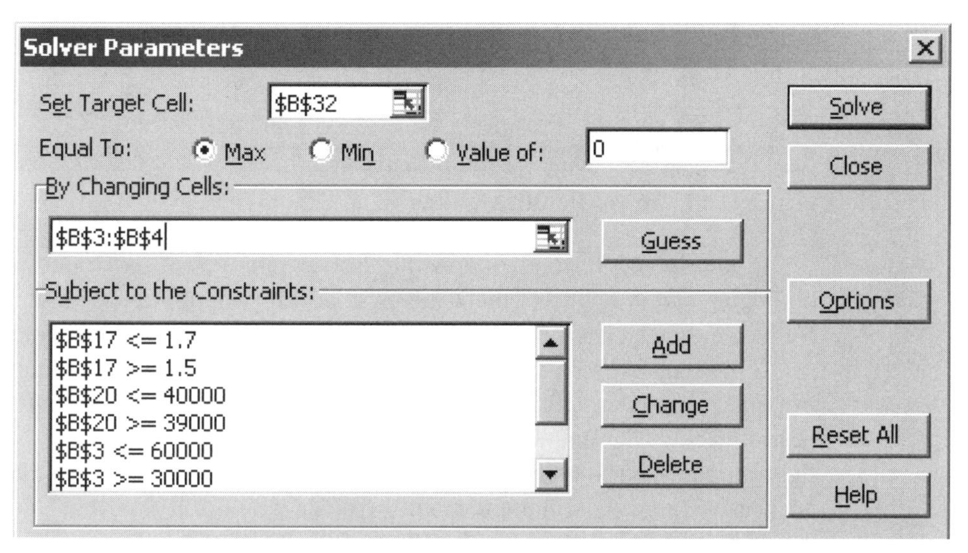

Figure D-12 A portion of the constraints entered in the Solver Parameters window

Using the scroll arrow, reveal the rest of the constraints, as shown in Figure D-13.

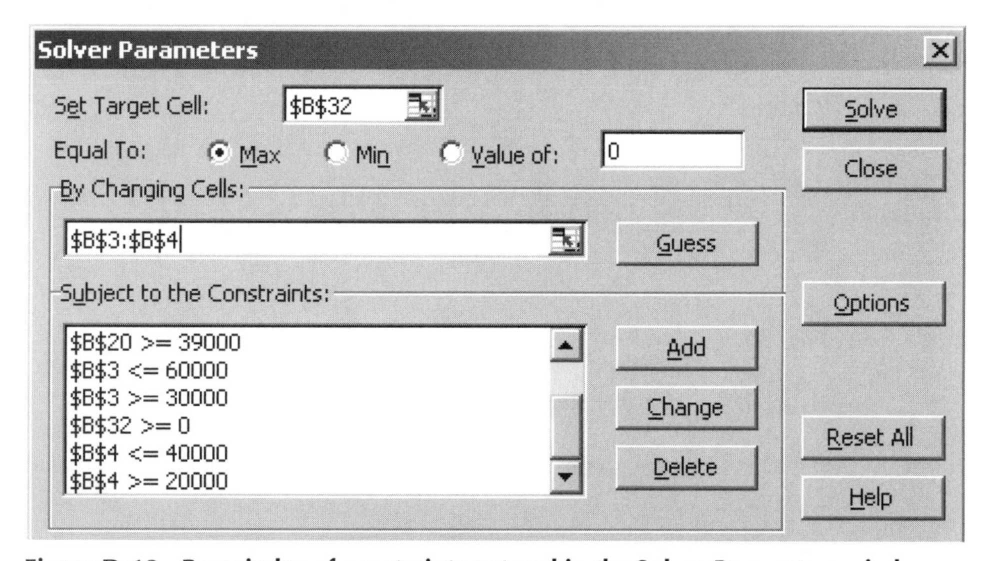

Figure D-13 Remainder of constraints entered in the Solver Parameters window

Computing the Solver's Answer

To have the Solver actually calculate answers, click Solve in the upper-right corner of the Solver Parameters window. The solver does its work in the background—you do not see the internal calculations. Then the Solver gives you a Solver Results window, as shown in Figure D-14.

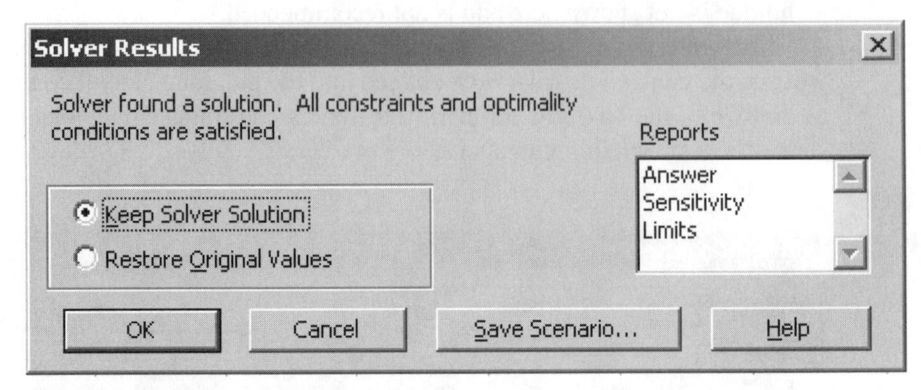

Figure D-14 Solver Results window

In the Solver Results window, the Solver tells you it has found a solution and that the optimality conditions were met. This is a very important message—you should always check for it. It means an answer was found and the constraints were satisfied.

By contrast, your constraints might be such that the Solver cannot find an answer. For example, suppose you had a constraint that said, in effect: "Net income must be at least a billion dollars." That amount cannot be reached, given so few basketballs and footballs and these prices. The Solver would report that no answer is feasible. The Solver may find an answer by ignoring some constraints. Solver would tell you that too. In either case, there would be something wrong with your model, and you would need to rework it.

There are two ways to see your answers. One way is to click OK. This lets you see the new changing cell values. A more formal (and complete) way is to click Answer in the Reports box, and then click OK. This puts detailed results into a new sheet in your Excel book. The new sheet is called an Answer Report. All answer reports are numbered sequentially as you run the Solver.

To see the Answer Report, click its tab, as shown in Figure D-15. (Here, this is Answer Report 1.)

26	B4	NUMBER OF FOOTBALLS	38095.2442	$B
27	B3	NUMBER OF BASKETBALLS	57142.85348	$B
28				
29				
30				
31				

Answer Report 1 / Sheet1 / Sheet2 / Sheet3

Figure D-15 Answer Report Sheet tab

This takes you to the Answer Report. The top portion of the report is shown in Figure D-16.

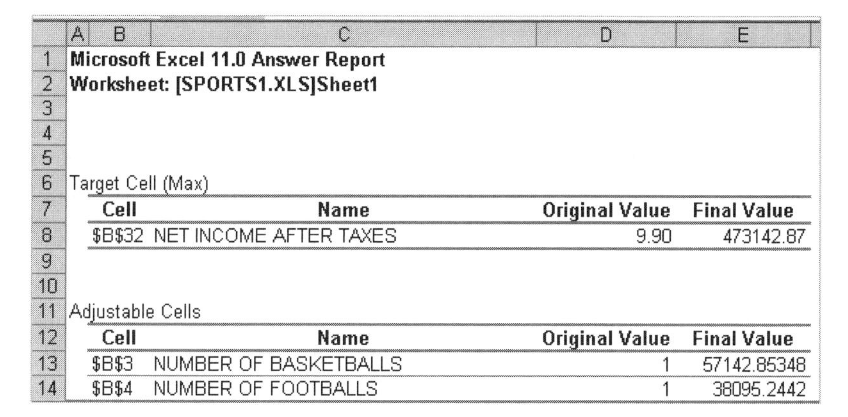

	A B	C	D	E
1	**Microsoft Excel 11.0 Answer Report**			
2	**Worksheet: [SPORTS1.XLS]Sheet1**			
3				
4				
5				
6	Target Cell (Max)			
7	**Cell**	**Name**	**Original Value**	**Final Value**
8	B32	NET INCOME AFTER TAXES	9.90	473142.87
9				
10				
11	Adjustable Cells			
12	**Cell**	**Name**	**Original Value**	**Final Value**
13	B3	NUMBER OF BASKETBALLS	1	57142.85348
14	B4	NUMBER OF FOOTBALLS	1	38095.2442

Figure D-16 Top portion of the Answer Report

Here is the remainder of the Answer Report, as shown in Figure D-17.

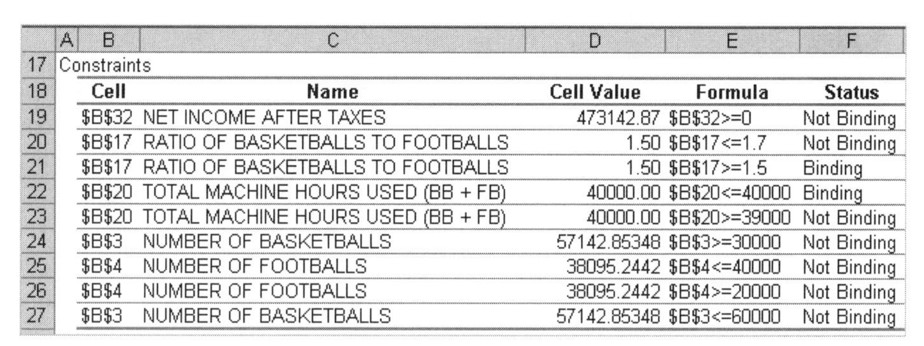

	A B	C	D	E	F
17	Constraints				
18	**Cell**	**Name**	**Cell Value**	**Formula**	**Status**
19	B32	NET INCOME AFTER TAXES	473142.87	B32>=0	Not Binding
20	B17	RATIO OF BASKETBALLS TO FOOTBALLS	1.50	B17<=1.7	Not Binding
21	B17	RATIO OF BASKETBALLS TO FOOTBALLS	1.50	B17>=1.5	Binding
22	B20	TOTAL MACHINE HOURS USED (BB + FB)	40000.00	B20<=40000	Binding
23	B20	TOTAL MACHINE HOURS USED (BB + FB)	40000.00	B20>=39000	Not Binding
24	B3	NUMBER OF BASKETBALLS	57142.85348	B3>=30000	Not Binding
25	B4	NUMBER OF FOOTBALLS	38095.2442	B4<=40000	Not Binding
26	B4	NUMBER OF FOOTBALLS	38095.2442	B4>=20000	Not Binding
27	B3	NUMBER OF BASKETBALLS	57142.85348	B3<=60000	Not Binding

Figure D-17 Remainder of Answer Report

At the beginning of this tutorial, the changing cells had a value of 1, and the income was $9.90 (Original Value). The optimal solution values (Final Value) are also shown: $473,142.87 for net income (the target), and 57,142.85 basketballs and 38,095.24 footballs for the changing (adjustable) cells. (Of course, you cannot make a part of a ball. The Solver can be asked to find only integer solutions; this technique is discussed at the end of this tutorial.)

The report also shows detail for the constraints: the constraint expression and the value that the variable has in the optimal solution. "Binding" means the final answer caused Solver to bump up against the constraint. For example, the maximum number of machine hours was 40,000, and that is the value Solver used in finding the answer.

"Not Binding" means the reverse. A better word for "binding" might be "constraining." For example, the 60,000 maximum basketball limit did not constrain the Solver.

The procedures used to change (edit) or delete a constraint are discussed later in this tutorial.

Print the worksheets (Answer Report and Sheet1). Save the Excel file (File—Save). Then, use File—Save As to make a new file called **SPORTS2.xls**, to be used in the next section of this tutorial.

Extending the Example

Next, you'll modify the sporting goods spreadsheet. Suppose that management wants to know what net income would be if certain constraints were changed. In other words, management wants to play "what-if" with certain Base Case constraints. The resulting second case is called the Extension Case. Let's look at some changes to the original Base Case conditions.

- Assume that maximum production constraints will be removed.
- Similarly, the basketball-to-football production ratios (1.5, 1.7) will be removed.
- There will still be minimum production constraints at some low level: Assume that at least 30,000 basketballs and 30,000 footballs will be produced.
- The machine-hours shared resource imposes the same limits as it did previously.
- A more ambitious profit goal is desired: The ratio of net income after taxes to total revenue should be greater than or equal to .33. This constraint will replace the constraint calling for profits greater than zero.

At the Keyboard

Begin by putting 1s in the changing cells. You will need to compute the ratio of net income after taxes to total revenue. Enter that formula in cell B21. (The formula should have the net income after taxes cell address in the numerator and the total revenue cell address in the denominator.) In the Extension Case, the value of this ratio for the Solver's optimal answer must be at least .33. Click the Add button and enter that constraint.

Then, in the Solver Parameters window, constraints that are no longer needed are highlighted (select by clicking) and deleted (click the Delete button). Do that for the net income >= 0 constraint, the maximum football and basketball constraints, and the basketball-to-football ratio constraints.

The minimum football constraint must be modified, not deleted. Select that constraint, then click Change. That takes you to the Add Constraint window. Edit the constraint so 30,000 is the lower boundary.

When you are finished with the constraints, your Solver Parameters window should look like the one shown in Figure D-18.

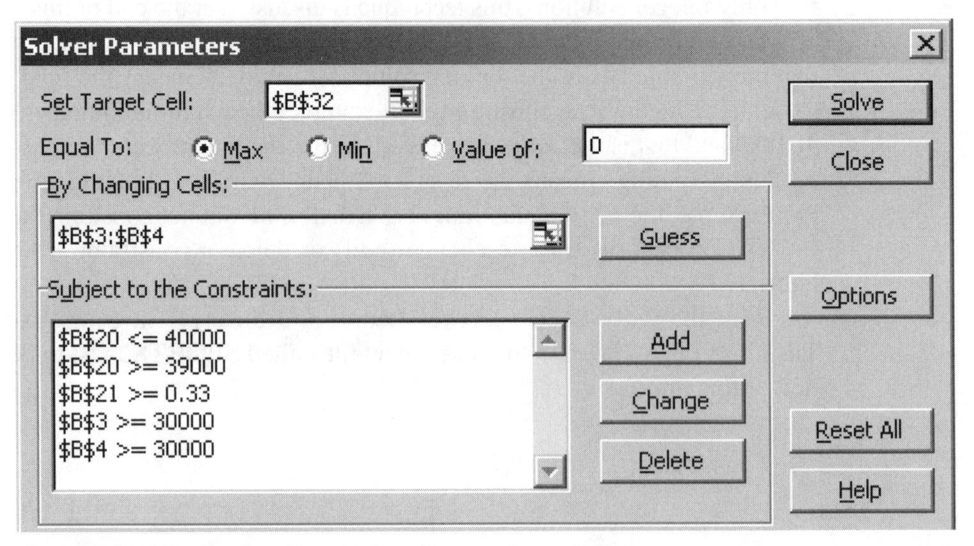

Figure D-18 Extension Case Solver Parameters window

You can tell Solver to solve for integer values. Here, cells B3 and B4 should be whole numbers. You use the Int constraint to do that. Figure D-19 shows entering the Int constraint.

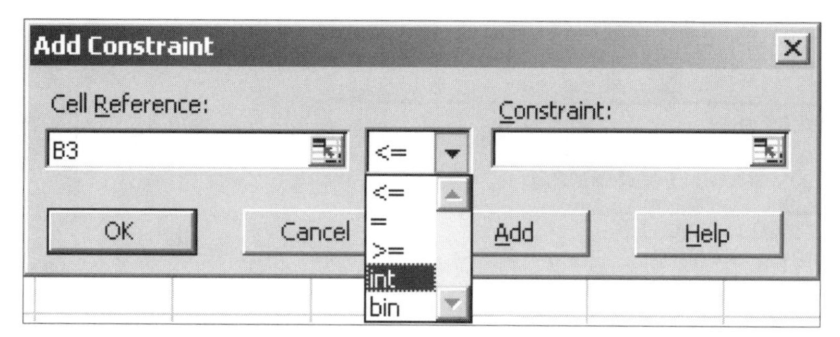

Figure D-19 Entering the Int constraint

Make those constraints for the changing cells. Your constraints should now look like the beginning portion of those shown in Figure D-20.

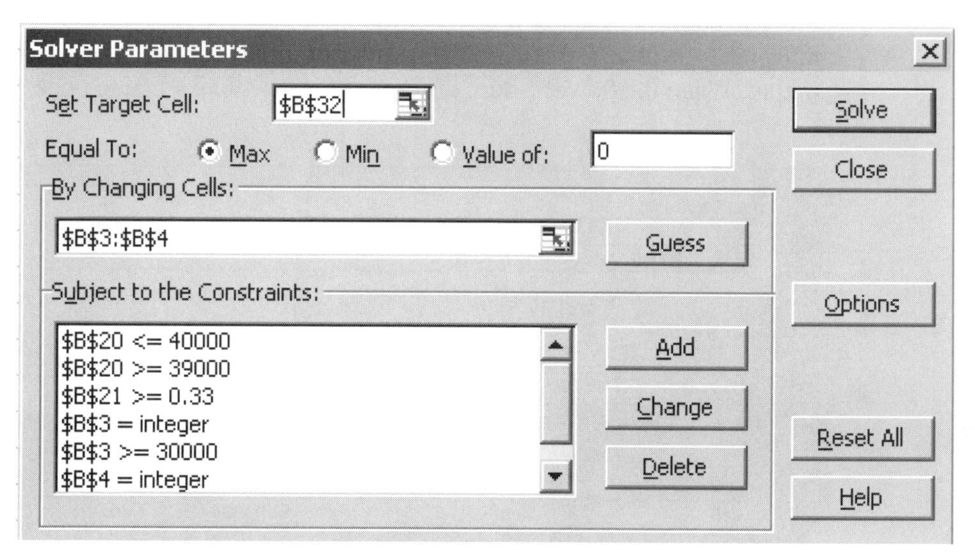

Figure D-20 Portion of Extension Case constraints

Scroll to see the remainder of the constraints, as shown in Figure D-21.

Tutorial D

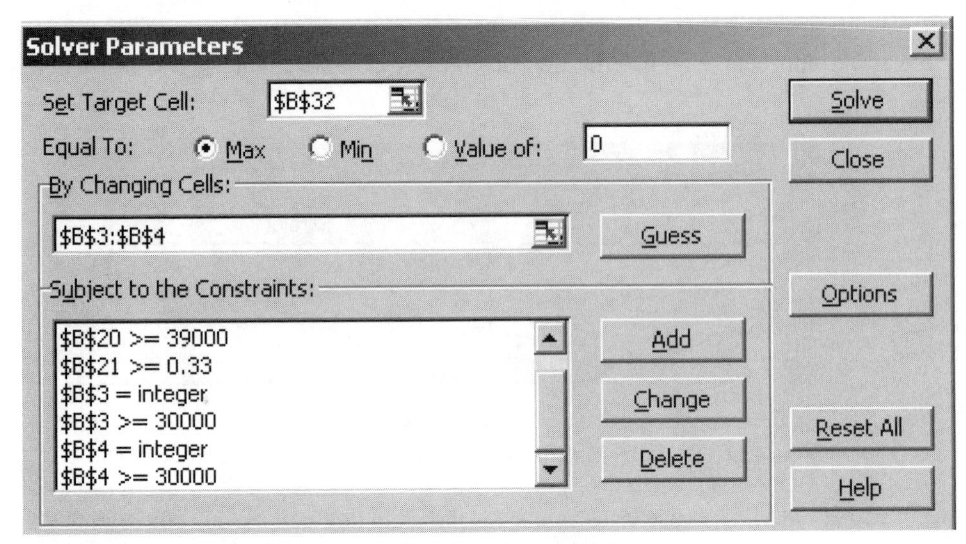

Figure D-21 Remainder of Extension Case constraints

The constraints are now only for the minimum production levels, the ratio of net income after taxes to total revenue, machine-hours shared resource constraints, and whole number output. When the Solver is run, the Answer Report looks like the one shown in Figure D-22.

	A	B	C	D	E	F
6	Target Cell (Max)					
7		Cell	Name	Original Value	Final Value	
8		B32	NET INCOME AFTER TAXES	9.90	556198.38	
9						
10						
11	Adjustable Cells					
12		Cell	Name	Original Value	Final Value	
13		B3	NUMBER OF BASKETBALLS	1	30000	
14		B4	NUMBER OF FOOTBALLS	1	83333	
15						
16						
17	Constraints					
18		Cell	Name	Cell Value	Formula	Status
19		B20	TOTAL MACHINE HOURS USED (BB + FB)	39999.90	B20<=40000	Not Binding
20		B20	TOTAL MACHINE HOURS USED (BB + FB)	39999.90	B20>=39000	Not Binding
21		B21	RATIO OF NET INCOME TO REVENUE	0.416109655	B21>=0.33	Not Binding
22		B3	NUMBER OF BASKETBALLS	30000	B3>=30000	Binding
23		B4	NUMBER OF FOOTBALLS	83333	B4>=30000	Not Binding
24		B3	NUMBER OF BASKETBALLS	30000	B3=integer	Binding
25		B4	NUMBER OF FOOTBALLS	83333	B4=integer	Binding

Figure D-22 Extension Case Answer Report

The Extension Case answer differs from the Base Case answer. Which production schedule should management use? The one that has maximum production limits? Or the one that has no such limits? These questions are posed to get you to think about the purpose of using a DSS program. Two scenarios, the Base Case and the Extension Case, were modeled in the Solver. The very different answers are shown in Figure D-23.

	Base Case	Extension Case
Basketballs	57,143	30,000
Footballs	38,095	83,333

Figure D-23 The Solver's answers for the two cases

Can you use this output alone to decide how many of each kind of ball to produce? No, you cannot. You must also refer to the "Target," which in this case is net income. Figure D-24 shows the answers with net income target data.

	Base Case	Extension Case
Basketballs	57,143	30,000
Footballs	38,095	83,333
Net Income	$473,143	$556,198

Figure D-24 The Solver's answers for the two cases—with target data

Viewed this way, the Extension Case production schedule looks better, because it gives you a higher target net income.

At this point, you should save the **SPORTS2.xls** file (File—Save) and then close it (File—Close).

Tutorial D

Using the Solver on a New Problem

Here is a short problem that will let you test what you have learned about the Excel Solver.

Setting Up the Spreadsheet

Assume that you run a shirt-manufacturing company. You have two products: (1) polo-style T-shirts, and (2) dress shirts with button-down collars. You must decide how many T-shirts and how many button-down shirts to make. Assume that you'll sell every shirt you make.

At the Keyboard

Open a file called **SHIRTS.xls**. Set up a Solver spreadsheet to handle this problem.

CHANGING CELLS Section

Your changing cells should look like those shown in Figure D-25.

	A	B
1	**SHIRT MANUFACTURING EXAMPLE**	
2	CHANGING CELLS	
3	NUMBER OF T-SHIRTS	1
4	NUMBER OF BUTTON-DOWN SHIRTS	1

Figure D-25 Shirt manufacturing changing cells

CONSTANTS Section

Your spreadsheet should contain the constants shown in Figure D-26. A discussion of constant cells (and some of your company's operations) follows the figure.

	A	B
6	CONSTANTS	
7	TAX RATE	0.28
8	SELLING PRICE: T-SHIRT	8.00
9	SELLING PRICE: BUTTON-DOWN SHIRT	36.00
10	VARIABLE COST TO MAKE: T-SHIRT	2.50
11	VARIABLE COST TO MAKE: BUTTON-DOWN SHIRT	14.00
12	COTTON USAGE (LBS): T-SHIRT	1.50
13	COTTON USAGE (LBS): BUTTON-DOWN SHIRT	2.50
14	TOTAL COTTON AVAILABLE (LBS)	13000000
15	BUTTONS PER T-SHIRT	3.00
16	BUTTONS PER BUTTON-DOWN SHIRT	12.00
17	TOTAL BUTTONS AVAILABLE	110000000

Figure D-26 Shirt manufacturing constants

- The TAX RATE is .28 on pre-tax income, but no taxes are paid on losses.
- SELLING PRICE: You sell polo-style T-shirts for $8 and button-down shirts for $36.
- VARIABLE COST TO MAKE: It costs $2.50 to make a T-shirt and $14 to make a button-down shirt. These variable costs are for machine-operator labor, cloth, buttons, and so forth.
- COTTON USAGE: Each polo T-shirt uses 1.5 pounds of cotton fabric. Each button-down shirt uses 2.5 pounds of cotton fabric.
- TOTAL COTTON AVAILABLE: You have only 13 million pounds of cotton on hand to be used to make all the T-shirts and button-down shirts.
- BUTTONS: Each polo T-shirt has 3 buttons. By contrast, each button-down shirt has 1 button on each collar tip, 8 buttons down the front, and 1 button on each cuff, for a total of 12 buttons. You have 110 million buttons on hand to be used to make all your shirts.

CALCULATIONS Section

Your spreadsheet should contain the calculations shown in Figure D-27.

	A	B
19	CALCULATIONS	
20	RATIO OF NET INCOME TO TOTAL REVENUE	
21	COTTON USED: T-SHIRTS	
22	COTTON USED: BUTTON-DOWN SHIRTS	
23	COTTON USED: TOTAL	
24	BUTTONS USED: T-SHIRTS	
25	BUTTONS USED: BUTTON-DOWN SHIRTS	
26	BUTTONS USED: TOTAL	
27	RATIO OF BUTTON-DOWNS TO T-SHIRTS	

Figure D-27 Shirt manufacturing calculations

Calculations (and related business constraints) are discussed next.

- RATIO OF NET INCOME TO TOTAL REVENUE: The minimum return on sales (ratio of net income after taxes divided by total revenue) is .20.
- COTTON USED/BUTTONS USED: You have a limited amount of cotton and buttons. The usage of each resource must be calculated, then used in constraints.
- RATIO OF BUTTON-DOWNS TO T-SHIRTS: You think you must make at least 2 million T-shirts and at least 2 million button-down shirts. You want to be known as a balanced shirtmaker, so you think that the ratio of button-downs to T-shirts should be no greater than 4:1. (Thus, if 9 million button-down shirts and 2 million T-shirts were produced, the ratio would be too high.)

INCOME STATEMENT Section

Your spreadsheet should have the income statement skeleton shown in Figure D-28.

	A	B
29	**INCOME STATEMENT**	
30	T-SHIRT REVENUE	
31	BUTTON-DOWN SHIRT REVENUE	
32	TOTAL REVENUE	
33	VARIABLE COSTS: T-SHIRTS	
34	VARIABLE COSTS: BUTTON-DOWNS	
35	TOTAL COSTS	
36	INCOME BEFORE TAXES	
37	INCOME TAX EXPENSE	
38	NET INCOME AFTER TAXES	

Figure D-28 Shirt manufacturing income statement line items

The Solver's target is net income, which must be maximized.

Use the table shown in Figure D-29 to write out your constraints before entering them into the Solver.

Expression in English	Fill in the Excel Expression
Net income to revenue	_____ >= _____
Ratio of BDs to Ts	_____ <= _____
Min T-shirts	_____ >= _____
Min button-downs	_____ >= _____
Usage of buttons	_____ <= _____
Usage of cotton	_____ <= _____

Figure D-29 Logic of shirt manufacturing constraints

When you are finished with the program, print the sheets. Then, use File—Save, File—Close, and then File—Exit, to leave Excel.

Trouble-Shooting the Solver

Use this section to overcome problems with the Solver and as a review of some Windows file-handling procedures.

Rerunning a Solver Model

Assume that you have changed your spreadsheet in some way and want to rerun the Solver to get a new set of answers. (For example, you may have changed a constraint or a formula in your spreadsheet.) Before you click Solve again to rerun the Solver, you should put the number 1 in the changing cells. The Solver can sometimes give odd answers if its point of departure is a set of prior answers.

Creating Over-Constrained Models

It is possible to set up a model that has no logical solution. For example, in the second version of the sporting goods problem, suppose that you had specified that at least 1 million basketballs were needed. When you clicked Solve, the Solver would have tried to compute an answer, but then would have admitted defeat by telling you that no feasible solution is possible, as shown in Figure D-30.

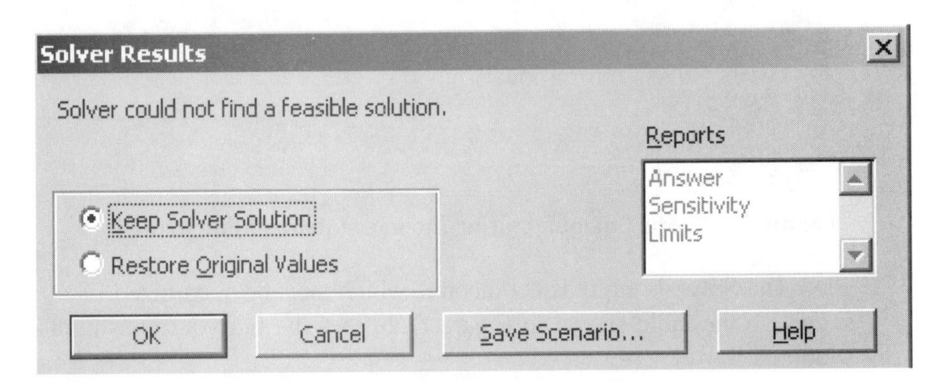

Figure D-30 Solver Results message: Solution not feasible

In the Reports window, the choices (Answer, etc.) would be in gray—indicating they are not available as options. Such a model is sometimes called "over-constrained."

Setting a Constraint to a Single Amount

It's possible you'll want an amount to be a specific number, as opposed to a number in a range. For example, if the number of basketballs needed to be exactly 30,000, then the "equals" operator would be selected, as shown in Figure D-31.

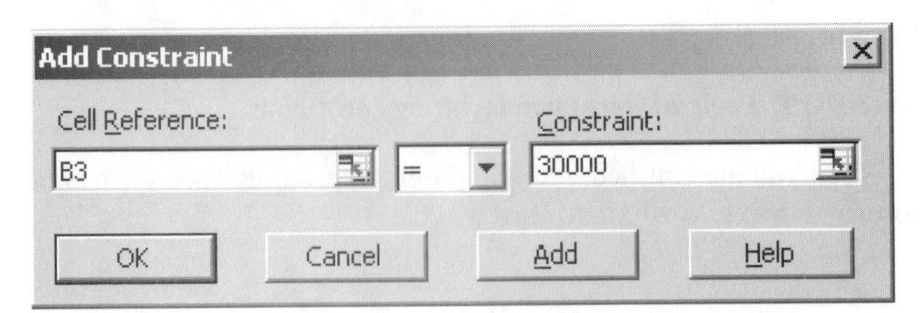

Figure D-31 Constraining a value to equal a specific amount

Setting a Changing Cell to an Integer

You may want to force changing cell values to be integers. The way to do that is to select the Int operator in the Add Constraint window. This was described in a prior section.

Forcing the Solver to find only integer solutions slows the Solver down. In some cases, the change in speed can be noticeable to the user. Doing this can also prevent the Solver from seeing a feasible solution—when one can be found if the Solver is allowed to find non-integer answers. For these reasons, it's usually best not to impose the integer constraint unless the logic of the problem demands it.

Deleting Extra Answer Sheets

Suppose that you've run different scenarios, each time asking for an Answer Report. As a result, you have a number of Answer Report sheets in your Excel file, but you don't want to keep them all. How do you get rid of an Answer Report sheet? Follow this procedure: First, get the unwanted Answer Report sheet on the screen by clicking the sheet's tab. Then select Edit—Delete Sheet. You will be asked if you really mean it. If you do, click accordingly.

Restarting the Solver with All-New Constraints

Suppose that you wanted to start over, with a new set of constraints. In the Solver Parameters window, click Reset All. You will be asked if you really mean it, as shown in Figure D-32.

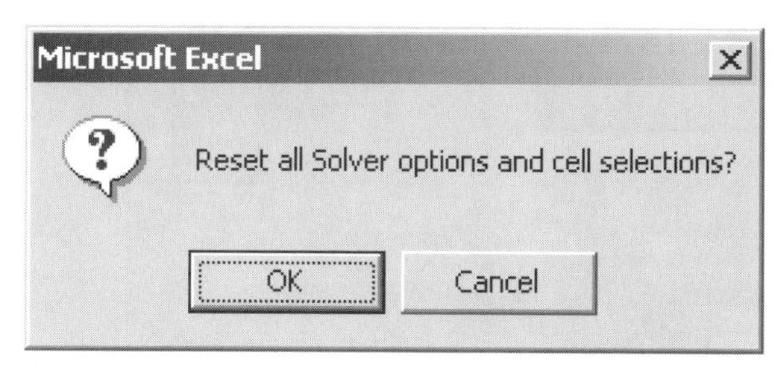

Figure D-32 Reset options warning query

If you do, then select OK. This gives you a clean slate, with all entries deleted, as shown in Figure D-33.

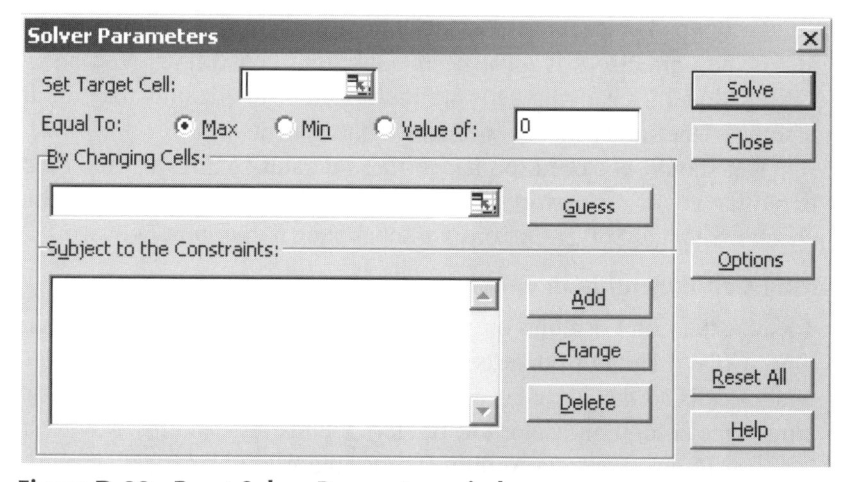

Figure D-33 Reset Solver Parameters window

As you can see, the target cell, changing cells, and constraints have been reset. From this point, you can specify a new model.

If you select Reset All, you really are starting over. If you merely want to add, delete, or edit a constraint, do not use Reset All. Use the Add, Delete, or Change buttons, as the case may be.

Solver Options Window

The Solver has a number of internal settings that govern its search for an optimal answer. If you click the Options button in the Solver Parameters window, you will see the defaults for these settings, as shown in Figure D-34:

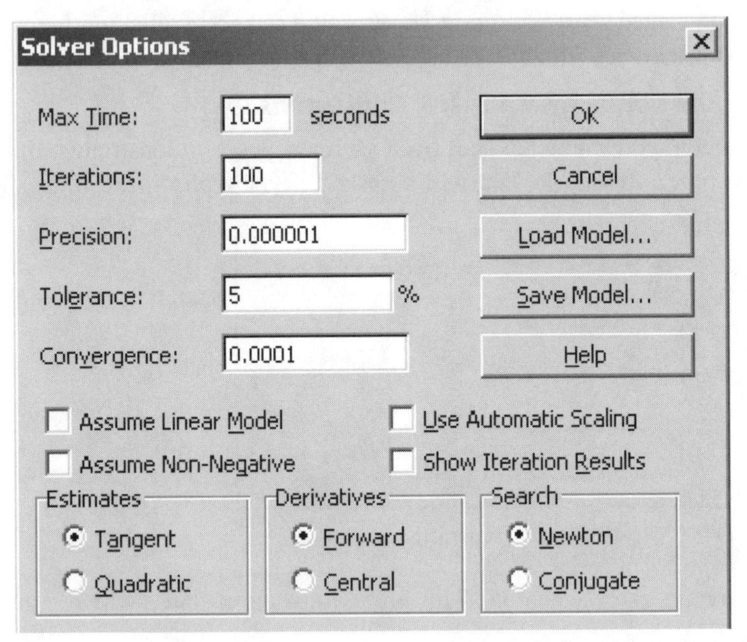

Figure D-34 Solver Options window, with default settings for Solver Parameters

Very broadly speaking, Solver Options govern how long the Solver works on a problem and/or how precisely it must be in satisfying constraints. You should not check Assume Linear Model if changing cells are multiplied or divided (as they are in this book's cases) or if some of the spreadsheet's formulas use exponents.

You should not need to change these default settings for the cases in this book. If you think that your Solver work is correct, but Solver cannot find a feasible solution, you should check to see that Solver Options are set as shown in Figure D-34.

Printing Cell Formulas in Excel

To show the Cell Formulas on the screen, press the Ctrl and the left quote (`) keys at the same time: Ctrl-`. (The left quote is usually on the same key as the tilde [~].) This automatically widens cells so the formulas can be read. You can change cell widths by clicking and dragging at the column indicator (A, B, etc.) boundaries. Another way to show the cells formulas is to use Tools—Formula Auditing—Formula Auditing Mode.

To print the formulas, just use File—Print. Print the sheet as you would normally. To restore the screen to its typical appearance (showing values, not formulas), press Ctrl-` again. (It's like a toggle switch.) If you did not change any column widths when in the cell formula view, the widths will be as they were.

Review of Printing, Saving, and Exiting Procedures

Print the Solver spreadsheets in the normal way. Activate the sheet, then select File—Print. You can print an Answer Report sheet in the same way.

To save a file, use File—Save, or File—Save As. Be sure to select Drive A: in the Drive window, if you intend your file to be on a diskette. When exiting from Excel, always start with File—Close (with the diskette in Drive A:), then select File—Exit. Only then should you take the diskette out of Drive A:.

NOTE

If you merely use File—Exit (not closing first), you risk losing your work.

Sometimes, you might think that the Solver has an odd sense of humor. For instance, your results might differ from the target answers that your instructor provides for a case. Thinking that you've done something wrong, you ask to compare your cell formulas and constraint expressions with those your instructor created. Lo and behold, you can see no differences! Surprisingly, the Solver can occasionally produce slightly different outputs from inputs that are seemingly the same, for no apparent reason. Perhaps, for your application, the order of the constraints matters, or even the order in which they are entered. In any case, if you are close to the target answers but cannot see any errors, it's best to see your instructor for guidance, rather than to spin your wheels.

Here is another example of the Solver's sense of humor. Assume that you ask for Integer changing cell outputs. The Solver may tell you that the correct output is 8.0000001, or 7.9999999. In such situations, the Solver is apparently just not sure about its own rounding! You merely humor the Solver and (continuing the example) take the result as the integer 8, which is what the Solver is trying to say in the first place.

CASE 9

The Baseball Park Configuration Decision

DECISION SUPPORT USING EXCEL

PREVIEW

In this case, you will use the Excel Solver to provide decision support to a construction company planning to build a baseball park. The company needs to know the optimum number of types of seats, from box seats to "cheap seats," to put in a new ballpark. You will quantify the financial impacts of different ballpark construction alternatives for the construction company.

PREPARATION

- Review spreadsheet concepts discussed in class and/or in your textbook. This case requires an understanding of these Excel functions: Round( ), IF( ), and Sum( ).
- Complete any exercises that your instructor assigns.
- Complete any part of Tutorial D that your instructor assigns, or refer to it as necessary.
- Review file-saving procedures for Windows programs. These are discussed in Tutorial C.
- Refer to Tutorial E as necessary.

BACKGROUND

Your hometown major league baseball team has decided to build a new baseball park. The land has been purchased, and a construction company, Field of Dreams Corporation, has been hired to build the park. Field of Dreams has hired you to provide decision support during the ballpark's design phase.

The team's current stadium was built for baseball and football. It has an enclosed oval shape, with seating around the entire circumference. When filled to capacity, the park can accommodate 70,000 people. Football crowds can fill the stadium to capacity. Baseball games, however, don't fill the stadium to capacity; usually, 30,000–40,000 fans attend each game. Crowds of more than 50,000 fans are rare.

The new park will be for baseball only. Unlike the old stadium, it will not be oval in shape, nor will it have seating all around its periphery. Rather, it will be **U**-shaped, like a horseshoe. Home plate would be at the closed end of the **U**. Seating would be all around the **U**, but there would be no seats at the open end.

Field of Dreams' engineers have calculated the length around the **U** (again, leaving out the open part) as 1375 feet from the left-field foul pole tip to the right-field foul pole tip. The engineers plan to build a stadium that has 100 rows of seats that extend back and up from the playing field.

In fact, the geometry of a stadium is such that the row closest to the playing field (row 1) is shorter in length than the row farthest from the playing field (row 100). However, some of those rows are broken by equipment, exits, and concessions, so to simplify calculations for this study, you can assume that each of the 100 rows is 1375 feet in length.

The construction engineers plan to install four kinds of seats:

- *Box Seats* are the rows of seats closest to the playing field and are usually considered the most desirable.
- *Red Seats* are the rows of seats in the next level up.
- *Blue Seats* are the rows of seats in the level above the red seats.
- *Yellow Seats* are the farthest rows up and away from the playing field. They are the least desirable seats because they are the farthest from the field.

Traditionally in baseball parks, the seats closest to the playing field cost the most, and the new park will follow that rule. Thus, the most well-to-do fans (or those who are truly fanatical about the game) buy the seats closest to the field, and those who are least well-to-do (or just do not want to pay very much for a ticket) sit farthest away. Admission ticket prices per game, by seat type, are shown in Figure 9-1.

Seat Type	Price per Game
Box Seat	$50
Red Seat	$35
Blue Seat	$25
Yellow Seat	$20

Figure 9-1 Ticket prices per seat type

The seat types differ by more than just paint color and their positions in the ballpark. The most-expensive seats will be the widest and, therefore, the most comfortable. The least-expensive seats will be the narrowest and, therefore, the least comfortable. The width of seats, by seat type, is shown in Figure 9-2.

Seat Type	Seat Width (in feet)
Box Seat	3.50
Red Seat	3.25
Blue Seat	3.00
Yellow Seat	2.75

Figure 9-2 Seat widths per seat type

Thus, because of seat widths, the number of seats in a Box Seat row will be fewer than the number of seats in a Red Seat row, and so on.

People who come to ballgames usually buy souvenirs and food. In fact, many fans spend more money buying souvenirs and food than they do buying their admission ticket. The team's research department has determined that there is a correlation between where fans sit and how much they spend on souvenirs and food. Figure 9-3 shows the average expenditures per game.

Seat Type	Spent on "Other" (per Game)
Box Seat	$30
Red Seat	$25
Blue Seat	$15
Yellow Seat	$10

Figure 9-3 Amounts spent on souvenirs and food per seat type

The new field will also need a parking lot that can accommodate 15,000 cars. The team's researchers have studied how fans get to the ballpark. They have found that less-affluent fans are more likely to use public transportation rather than drive a car to the game. Their research reveals the following averages:

- For every 2 fans sitting in Box Seats, 1 car arrives at the ballpark.
- For every 3 fans sitting in Red Seats, 1 car arrives at the ballpark.
- For every 3 fans sitting in Blue Seats, 1 car arrives at the ballpark.
- For every 5 fans sitting in Yellow Seats, 1 car arrives at the ballpark.

Major league baseball rules say a stadium must be able to seat at least 44,000 fans, so, at a minimum, the park must be able to seat that many fans. However, team management wants the park to seem "intimate," so they do not want to be able to seat more than 46,000 fans.

Team management also wants a balance among the different seating types. The minimum number of rows for each kind of seat is shown in Figure 9-4.

Seat Type	Minimum Number of Rows
Box Seat	10
Red Seat	10
Blue Seat	10
Yellow Seat	10

Figure 9-4 Minimum number of rows per seat type

The maximum number of rows for each kind of seat is shown in Figure 9-5.

Seat Type	Maximum Number of Rows
Box Seat	30
Red Seat	40
Blue Seat	40
Yellow Seat	50

Figure 9-5 Maximum number of rows per seat type

Field of Dreams wants you to answer this question: How many rows of each type of seat should be installed? You must model two situations: a Base Case and an Extension Case. You will use the Solver as your decision-support modeling tool. After you have finished the Base Case, you will modify its spreadsheet to create the Extension Case, which will let you play "what if" with the Base Case. Field of Dreams will use your input to decide how to configure the new ballpark. Finally, you will write a memorandum that recommends a course of action. Your memorandum will include (as attachments) printouts of the Solver spreadsheets and reports documenting each model. You will give your instructor your memo, printouts, and your disk.

Assignment 1 Creating a Spreadsheet for Decision Support

In this assignment, you will produce a spreadsheet that models the business decision. In Assignment 1A, you will make a Solver spreadsheet to model the configuration decision's Base Case. In Assignment 1B, you will make a Solver spreadsheet to model the configuration's Extension Case. Then, in Assignment 2, you will use the spreadsheet models to develop information needed to recommend the best configuration for the ballpark, and then you will document your recommendation in a memorandum to Field of Dreams and to the team's management. In Assignment 3, you will give your recommendation in an oral presentation.

Next, you will create the spreadsheet models of the configuration decision. Your spreadsheet should have the following sections:

- **CHANGING CELLS**
- **CONSTANTS**
- **CALCULATIONS**
- **REVENUE**

Your spreadsheet will also include the decision constraints. *The spreadsheet skeleton is available to you, so you need not type in the skeleton if you do not want to do so.* To access the spreadsheet skeleton for the Base Case, go to your data files. Select Case 9, and then select **PARK1.xls**. For the Extension Case, you will use the **PARK2.xls** data file.

Assignment 1A: Creating the Spreadsheet—Base Case

You will build a model of the stadium construction problem previously described. Your model, when run, will reveal how many rows of each type of seat to install. Of course, the total number of rows must be 100. **Team management's goal is to maximize total revenue, subject to the various constraints.** For the purpose of this study, assume that all seats in the ballpark will be sold, and determine total revenue for a one-game sellout crowd.

Team management does not want the new stadium to have a massive "cheap seats" area. So, their rule is this: *The number of rows of Red Seats plus the number of rows of Blue Seats, combined, must be greater than or equal to the number of rows of Yellow Seats.*

A discussion of each spreadsheet section follows. The discussion is about (1) how each section should be set up, and (2) the logic of the formulas in the section's cells.

CHANGING CELLS Section

Your spreadsheet should have the changing cells shown in Figure 9-6.

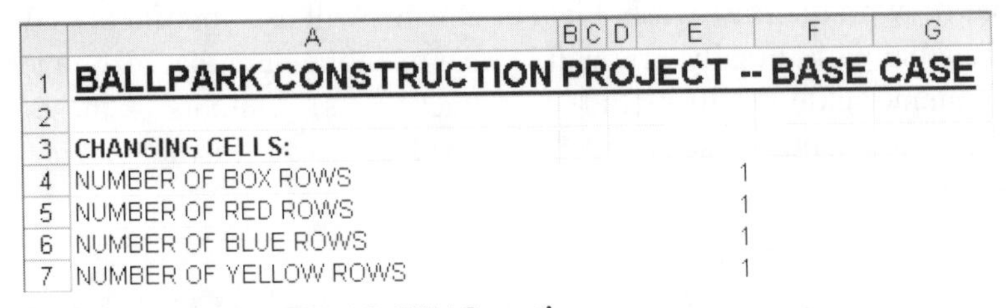

	A	B	C	D	E	F	G
1	BALLPARK CONSTRUCTION PROJECT -- BASE CASE						
2							
3	CHANGING CELLS:						
4	NUMBER OF BOX ROWS				1		
5	NUMBER OF RED ROWS				1		
6	NUMBER OF BLUE ROWS				1		
7	NUMBER OF YELLOW ROWS				1		

Figure 9-6 CHANGING CELLS section

You are asking the Solver model to compute the number of rows for each type of seat. Start with a *1* in each cell. The Solver will change each *1* as it computes the answer. Do not allow the Solver to recommend a fraction of a row.

CONSTANTS Section

Your spreadsheet should have the constants shown in Figure 9-7. An explanation of the line items follows the figure.

	A	B	C	D	E	F	G
9	**CONSTANTS:**						
10	NUMBER OF FEET IN A ROW				1375		
11	SEAT WIDTH: BOX				3.50		
12	SEAT WIDTH: RED				3.25		
13	SEAT WIDTH: BLUE				3.00		
14	SEAT WIDTH: YELLOW				2.75		
15	NUMBER OF ROWS				100		
16	PRICE PER TICKET: BOX				50		
17	PRICE PER TICKET: RED				35		
18	PRICE PER TICKET: BLUE				25		
19	PRICE PER TICKET: YELLOW				20		
20	AVG OTHER PURCHASES: BOX				30		
21	AVG OTHER PURCHASES: RED				25		
22	AVG OTHER PURCHASES: BLUE				15		
23	AVG OTHER PURCHASES: YELLOW				10		
24	MIN # ROWS: BOX				10		
25	MIN # ROWS: RED				10		
26	MIN # ROWS: BLUE				10		
27	MIN # ROWS: YELLOW				10		
28	MAX # ROWS:BOX				30		
29	MAX # ROWS:RED				40		
30	MAX # ROWS:BLUE				40		
31	MAX # ROWS:YELLOW				50		
32	NUMBER OF PEOPLE PER CAR:BOX				2		
33	NUMBER OF PEOPLE PER CAR:RED				3		
34	NUMBER OF PEOPLE PER CAR:BLUE				3		
35	NUMBER OF PEOPLE PER CAR:YELLOW				5		
36	MIN ATTENDANCE				44000		
37	MAX ATTENDANCE				46000		
38	NUMBER OF PARKING PLACES				15000		

Figure 9-7 CONSTANTS section

- NUMBER OF FEET IN A ROW: Each row is 1375 feet in length, as stated.
- SEAT WIDTH (BOX, RED, BLUE, YELLOW): Each seat type has a specific width (Figure 9-2).
- NUMBER OF ROWS: The ballpark must have exactly 100 rows of seats, as stated.
- PRICE PER TICKET (BOX, RED, BLUE, YELLOW): Admission tickets for each seat type sell for a specific amount per game (Figure 9-1).
- AVERAGE OTHER PURCHASES (BOX, RED, BLUE, YELLOW): People in each of the seating types spend an average sum on food and souvenirs (Figure 9-3).
- MIN # ROWS (BOX, RED, BLUE, YELLOW): A minimum number of rows for each type of seat that must be installed (Figure 9-4).

- MAX # ROWS (BOX, RED, BLUE, YELLOW): A maximum number of rows for each type of seat that can be installed (Figure 9-5).
- NUMBER OF PEOPLE PER CAR (BOX, RED, BLUE, YELLOW): The number of people coming to the ballpark by car corresponds to each seat type, as stated.
- MIN ATTENDANCE: Major League rules state that the stadium must have a minimum of 44,000 seats.
- MAX ATTENDANCE: Team management has set a maximum of 46,000 for the total number of seats.
- NUMBER OF PARKING PLACES: The parking lot must accommodate 15,000 vehicles, as stated.

CALCULATIONS Section

Your spreadsheet should calculate the amounts shown in Figure 9-8. They will be used in the REVENUE section and/or in the CONSTRAINTS section. Calculations are based on values in the CHANGING CELLS section and/or in the CONSTANTS section and/or on other calculated values. An explanation of the line items follows the figure.

	A	B	C	D	E	F	G
41	**CALCULATIONS:**						
42	NUMBER OF PEOPLE PER ROW-BOX						
43	NUMBER OF PEOPLE PER ROW-RED						
44	NUMBER OF PEOPLE PER ROW-BLUE						
45	NUMBER OF PEOPLE PER ROW-YELLOW						
46	NUMBER OF BOX SEATS						
47	NUMBER OF RED SEATS						
48	NUMBER OF BLUE SEATS						
49	NUMBER OF YELLOW SEATS						
50	TOTAL NUMBER OF SEATS						
51	NUMBER CARS - BOX						
52	NUMBER CARS - RED						
53	NUMBER CARS - BLUE						
54	NUMBER CARS - YELLOW						
55	TOTAL NUMBER OF CARS						
56	TOTAL NUMBER OF ROWS						
57	NUMBER OF RED + BLUE						

Figure 9-8 CALCULATIONS section

- NUMBER OF PEOPLE PER ROW (BOX, RED, BLUE, YELLOW): Each row and its seat type can contain a specific number of people. The calculation is a function of the length of the row and of the width of seats in the row. You should not compute a fraction of a person—use the Round() function (zero decimals) to prevent fractions. Calculate the number of people per row for each seat type.
- NUMBER OF SEATS (BOX, RED, BLUE, YELLOW): For each seat type, this is a function of the total number of seat rows and the number of seats in each row. Calculate the number of seats for each seat type.
- TOTAL NUMBER OF SEATS: This is the total number of seats of all seat types in the ballpark.

- NUMBER OF CARS (BOX, RED, BLUE, YELLOW): This is a function of the number of fans per car for each seat type and of the number of each seat type sold. Use the Round() function to prevent fractions of a car. Calculate the number of cars for each seat type.
- TOTAL NUMBER OF CARS: Compute the total number of cars that will be needed to transport fans in all the seating types.
- TOTAL NUMBER OF ROWS: Compute the total number of rows to be installed in the stadium.
- NUMBER OF RED + BLUE: Compute the total number of rows installed for Red and Blue seating.

REVENUE Section

Compute total revenues for each kind of seat, as shown in Figure 9-9. Line items are discussed after the figure.

	A	B	C	D	E	F	G
61	**REVENUE:**						
62	TICKET SALES: BOX						
63	TICKET SALES: RED						
64	TICKET SALES: BLUE						
65	TICKET SALES: YELLOW						
66	TOTAL TICKET REVENUE						
67	OTHER REVENUE: BOX						
68	OTHER REVENUE: RED						
69	OTHER REVENUE: BLUE						
70	OTHER REVENUE: YELLOW						
71	TOTAL OTHER REVENUE						
72	TOTAL REVENUE						

Figure 9-9 REVENUE section

- TICKET SALES (BOX, RED, BLUE, YELLOW): For each seat type, this is a function of the number of seats installed and the price of a seat ticket.
- TOTAL TICKET REVENUE: Compute total ticket revenue for all types of seats.
- OTHER REVENUE (BOX, RED, BLUE, YELLOW): For each seat type, this is a function of the number of seats installed and corresponding amount spent on food and souvenirs.
- TOTAL OTHER REVENUE: Compute total other revenue for all kinds of seats.
- TOTAL REVENUE: Compute the total revenue for both ticket sales and "other" sales.

CONSTRAINTS and Running the Solver

Determine the constraints. Enter the Base Case decision constraints, using the Solver. Run the Solver. Make an Answer Report when the Solver says that a solution has been found that satisfies the constraints.

When you are finished, print the entire workbook, including the Answer Report. Save the Base Case Solver spreadsheet file (File—Save). For the Extension Case, use spreadsheet skeleton **PARK2.xls** in your data files.

Assignment 1B: Creating the Spreadsheet—Extension Case

Some in team management think that baseball fans should be given a more "friendly" experience at the ballpark. They want a much wider "concourse" area around the circumference of the stadium where fans can walk, get food to eat, play games, and mingle more freely. Management envisions more restaurants serving better food, for which the team would charge somewhat higher prices than in the Base Case. There would also be various kinds of games and amusements; for example, fans could bat against a pitching machine enclosed in netting. Or, fans could get their pictures taken with the team mascot.

This concourse area would be much more extensive than the concourse area planned for the Base Case configuration. The total area taken up by the ballpark cannot be increased, however. To make room for the larger concourse, space must be taken from stadium seating. The average length of each row would need to fall to 1300 feet. In effect, the seating **U** would be smaller, and the area around the **U** would be larger.

Team management thinks that fans will pay more for a wider variety of better food. In addition, revenue from selling souvenirs and other amusements would increase. Losing seating area might be a bad thing, but perhaps what is lost in seating will be more than made up in "other" revenue. The team's research department forecasts that per-game expenditures would be as shown in Figure 9-10:

Seat Type	Spent on "Other" per Game
Box Seat	$35
Red Seat	$30
Blue Seat	$25
Yellow Seat	$20

Figure 9-10 Amounts spent on other products—Extension Case

Those in management who argue for a larger concourse also feel that the rule about the relationship of Red Seats and Blue Seats to Yellow Seats can be jettisoned—if those who would sit in "cheap" seats far outnumber others, so be it. These managers argue that one fan's money is as good as another's in the expanded concourse area.

Modify the Extension Case spreadsheet (PARK2.XLS) to reflect this alternative. Run the Solver. Ask for an Answer Report when the Solver says that a solution has been found that satisfies the constraints. When you are done, print the entire workbook, including the Solver Answer Report. Save the workbook when you are finished. Close the file and exit Excel.

Assignment 2 Using the Spreadsheet for Decision Support

You have built the Base Case and the Extension Case models because you want to know the ballpark configuration in each scenario and which scenario yields the highest total revenue. You will now complete the case by (1) using the worksheets and Answer Reports to gather the data needed to decide which configuration to use, and (2) documenting your recommendation in a memorandum and (if your instructor specifies) an oral presentation.

Assignment 2A: Using the Spreadsheet to Gather Data

You have printed the spreadsheet and Answer Report sheet for each scenario. You can see how many rows of each kind of seating to install and revenue earned for each case. Note that data for each case on a separate sheet of paper for inclusion in a table in your memorandum. The form of your table is shown in Figure 9-11.

	Base Case	Extension Case
Number of rows of Box Seats		
Number of rows of Red Seats		
Number of rows of Blue Seats		
Number of rows of Yellow Seats		
Total number of seats in park		
Number of cars in parking lot		
Total other revenue		
Total revenue		

Figure 9-11 Form for table in memorandum

Other revenue will not appear in the Answer Reports because the value was not used in any constraints or in changing cells. Look in the REVENUE section of your worksheets for the other revenue values.

Assignment 2B: Documenting Your Recommendation in a Memorandum

Write a brief memorandum to Field of Dreams management and team management about the results. Observe the following requirements:

1. Your memorandum should have a proper heading (DATE / TO / FROM / SUBJECT). You may wish to use a Word memo template (**File**—click **New**, click **On my computer** in the Templates section, click the **Memos** tab, choose **Contemporary Memo**, then click **OK**).
2. Briefly outline the situation and the decisions to be made. State which configuration gives the higher total revenue—you would recommend that configuration. Team management wants to know if other revenue makes a difference in the recommendation. You should say whether it is the decisive factor in your recommendation.
3. Support your recommendation graphically by making a summary table in Word after the first paragraph such as that shown in Figure 9-11.

Enter a table into Word, using the following procedure:

1. Select the **Table** menu option, point to **Insert**, then click **Table**.
2. Enter the number of rows and columns.
3. Select **AutoFormat** and choose **Table Grid 1**.
4. Select **OK**, and then select **OK** again.

ASSIGNMENT 3 GIVING AN ORAL PRESENTATION

The Field of Dreams Corporation and team management so are impressed by your analysis, they have asked you to give a presentation explaining your recommendation to a wider group of team management and employees. Prepare to explain your analysis and recommendation to the group in 10 minutes or fewer. Use appropriate visual aids or handouts. Tutorial E has guidance on how to prepare and give an oral presentation.

DELIVERABLES

Assemble the following deliverables for your instructor:

1. Printout of your memorandum
2. Spreadsheet printouts
3. Disk or CD, which should have your Word memorandum file and your Excel spreadsheet files

Staple the printouts together, with the memorandum on top. If there are other .xls files on your disk or CD, write your instructor a note, stating the names of your Base Case and Extension Case .xls files.

Hoobah's Casino Revenue Maximization Problem

10
CASE

DECISION SUPPORT USING EXCEL

PREVIEW

In this problem, you will use the Excel Solver to give decision support to a company that has recently purchased a hotel with a ground-floor casino. Management must decide how to remodel the casino to maximize casino income. You will quantify the financial impacts of different alternatives.

PREPARATION

- Review spreadsheet concepts discussed in class and/or in your textbook. This case requires an understanding of the IF() and Sum() Excel functions.
- Complete any exercises that your instructor assigns.
- Complete any part of Tutorial D that your instructor assigns, or refer to them as necessary.
- Review file-saving procedures for Windows programs. These are discussed in Tutorial C.
- Refer to Tutorial E as necessary.

Background

The Hoobah Entertainment Corporation (HEC) owns many businesses in the entertainment industry. HEC has just purchased a hotel that can accommodate 1080 guests. It has a large first-floor area that will be devoted to a casino, dining, and stage shows.

HEC has an excellent marketing department and extensive experience in the hotel and gaming industries. Thus, management is certain that they can book the hotel rooms to full capacity each day. Their goal is to maximize first-floor revenue, and to do that, they want your help to target a mix of guests that will be most profitable.

Hotels with casinos attract a variety of guests—not just high-roller gamblers. These are the kinds of guests that HEC would like to target:

- *Machine players*—These gamblers play slot machines and poker machines.
- *Game-of-chance players*—These gamblers enjoy pitting their luck and skill against the odds. Such players like to play blackjack, roulette, and craps.
- *High rollers*—These gamblers play different kinds of card games and bet very large sums of money on each hand.
- *Entertainment-oriented guests*—These "showtime" guests go to the casino to have dinner and watch a stage show—and to watch the gamblers!

Each type of guest spends money differently and must be catered to differently. Thus, the income generated and the expenses incurred vary by type of guest. The HEC Marketing Department knows how to identify—and appeal to—each kind of guest. Marketing's task is to fill the hotel each day with a mix of guests that will maximize income for the business operation on the hotel's ground floor.

Of course, people other than guests can use the casino, and management hopes that people will come in off the street to use it.

The hotel's ground floor has 50,000 square feet of space. Not all of the space will be devoted to gaming, however. At one end of the floor, 5000 square feet will be devoted to a buffet dining area, where gamblers can eat. At the other end of the floor, a total of 5000 square feet would be devoted to a stage and a sit-down fine-dining area where diners could eat a meal while watching a show. The city's Fire Marshal says that no more than 1850 people can occupy the casino at one time. Thus, the number of hotel occupants in the casino and those who come in from the street cannot exceed 1850.

Guest "Base Revenue" per Day

Each kind of guest generates revenue differently. The hotel managers call this the "base revenue per day." Each type of gambling guest loses a certain average amount per day, and their losses are revenue to the hotel. By contrast, showtime guests generate show-ticket revenue. Base revenues per day are summarized in Figure 10-1.

Type of Guest	Base Revenue per Day
Machine player	$200
Game-of-chance player	$300
Showtime guest	$100
High roller	$5000

Figure 10-1 Base revenue per day by type of guest

Example: A machine player loses, on average, $200 to the casino each day.

Guest Food and Drink Revenue per Day

In addition, each type of guest generates a different amount of food and drink revenue: Slot and poker machine players usually "get into a zone" and do not eat very often or very much. By contrast, game-of-chance players usually enjoy taking a break to enjoy meals. Showtime guests generate the most food and drink revenue because they order expensive wines and entrées off a menu. High rollers generate no revenue because they are "comped"—that is, the hotel management provides complimentary meals and drinks to entice these big spenders to come to the casino. Food revenues per day are summarized in Figure 10-2.

Type of Guest	Food (and Drink) Revenues per Day
Machine player	$25
Game-of-chance player	$100
Showtime guest	$125
High roller	0

Figure 10-2 Food revenue per day by type of guest

Example: A machine player consumes on average $25 worth of food and drink each day in the casino.

Guest Expense per Day

Each type of guest generates different kinds of expenses. For example, machine poker players can be untidy, and extra janitorial staff must be hired to sweep up around these guests. This cost is quite low. Hosting game-of-chance players costs more: Usually, five or six game-of-chance players sit at a gaming table, and a dealer is required to deal and wait staff are required to bring drinks. Showtime guests require a waiter, bus boys, and so forth. High rollers are the most expensive. One dealer and cocktail server might be required for each high roller. In addition, extra security officers must patrol the high-roller area to insure the safety of these guests and to keep curious other patrons at bay. Expenses per day per guest are summarized in Figure 10-3.

Type of Guest	Expenses per Day
Machine player	1
Game-of-chance player	40
Showtime guest	10
High roller	500

Figure 10-3 Expense per day by type of guest

Example: The casino spends $40 each day to cater to the average game-of-chance player. If there were 100 such guests in the hotel that day, $4000 in expenses would be incurred.

Guest Number Minimum and Maximum per Day

Based on experience, management knows that it is not good to have too many of any one type of guest. A mix is needed. Each type of guest is stimulated by seeing a varied clientele. Thus, management has set minimums and maximums for each type of guest, as shown in Figure 10-4.

Type of Guest	Minimum Each Day	Maximum Each Day
Machine player	600	800
Game-of-chance player	400	800
Showtime guest	300	800
High roller	20	60

Figure 10-4 Minimum and maximum number of each type of guest

Example: On any day, there should be at least 600 machine player guests and no more than 800 machine player guests.

Guest Floor Space Required

Each kind of guest requires a certain amount of casino floor space. For example, for a machine player, there must be space for the machine and for the player to stand in front of it. A game-of-chance player requires room for a chair, and a portion of the areas taken up by the table and the dealer's space. A showtime guest needs room for a chair and part of a table where they are eating. High rollers require the most personal space for themselves, their dealer, and their privacy. Square footage per type of guest is detailed in Figure 10-5.

Type of Guest	Square Footage Needed
Machine player	15 square feet
Game-of-chance player	30 square feet
Showtime guest	10 square feet
High roller	100 square feet

Figure 10-5 Square footage needed per type of guest

Example: A slot machine player needs 15 square feet to gamble.

Guests' Influences on Other Guests

Oddly, different types of guests are stimulated by some types of guests and repulsed by others. This affects a guest's propensity to gamble and spend. HEC psychologists have quantified these effects quite precisely. Their observations are stated in Figure 10-6.

Type of Guest	Impact of Other Guests on Propensity to Spend
Machine player	These guests are not affected by any other type of guest. They just play!
Game-of-chance player	Two impacts: (1) For each 25 machine players on the floor, each player will spend $5 less. (2) For each 10 high rollers, $5 more will be lost.
Showtime guest	Dinner guests are distracted by machine players. For each 25 machine players, a showtime guest will spend $2 less on dinner and shows.
High roller	Two impacts: (1) These gamblers consider machine players to be amateurs, and they spend $5 less for each 25 machine players on the floor. (2) They are energized by game-of-chance players, however, and spend $10 more for each 25 game-of-chance players on the floor.

Figure 10-6 Rules governing adjustments to revenue by impact of other players

Example: There are 600 machine players on the floor. Each showtime dinner guest spends $2 * (600 /25) = $48 less for dinner.

An Alternative Viewpoint on the Guest Mix

Some members of management question the way dining areas are proposed. In the Base Case previously described, plans call for a sit-down dining area near the stage. Also, there would be a second dining area on the other side of the floor, where gamblers could order off the menu or eat a buffet. The long buffet tables take up quite a bit of floor space. Some members of management think the buffet could be removed. This would free up some space for more gamblers, without (in their view) changing the amount spent on food by gamblers: 1500 square feet would be gained in this way.

These managers also think that it is a mistake to have maximum numbers of players for game-of-chance players and for high rollers—in their view, any number of such players would be acceptable. This alternative way of doing things is called the Extension Case.

You must model two situations: the Base Case and the Extension Case. You will use the Solver as your decision-support modeling tool. Once you finish the Base Case, you modify the Base Case spreadsheet for the Extension Case. The Extension Case can be considered a way of playing "what if" with the Base Case. Finally, you will write a memo that recommends a course of action: the Base Case or Extension Case. Your memo will have as attachments the printouts of the Solver spreadsheets and reports documenting each model. You will turn in your memorandum, printouts, and your disk.

ASSIGNMENT 1 CREATING A SPREADSHEET FOR DECISION SUPPORT

In this assignment, you will produce a spreadsheet that models the business decision. In Assignment 1A, you will make a Solver spreadsheet to model the guest mix decision's Base Case. In Assignment 1B, you will make a Solver spreadsheet to model the Extension Case.

Then, in Assignments 2 and 3, you will use the spreadsheet models to develop information needed to recommend the best guest mix for the hotel. In Assignment 2, you will document your recommendation in a memorandum to HEC management. In Assignment 3, you will give your recommendation in an oral presentation.

Now you will create the spreadsheet model of the Base Case configuration decision. Your spreadsheet should have the sections that follow:

- **CHANGING CELLS**
- **CONSTANTS**
- **CALCULATIONS**
- **INCOME STATEMENT**

Your spreadsheets will also include the decision constraints. *The spreadsheet skeleton is available to you, so you need not type in the skeleton if you do not want to do so.* To obtain the spreadsheet skeletons for the Base Case, go to your Data Files, Case 10, and load **FLOOR1.xls**. For the Extension Case, you will need **FLOOR2.xls.**

Assignment 1A: Creating the Spreadsheet—Base Case

Build a model of the guest mix problem. Your model, when run, will tell HEC how many of each kind of casino guest to attract each day and, in turn, how to set up the casino floor. HEC management's goal is to maximize each day's casino income—total revenue less expenses, subject to the various constraints.

A discussion of each spreadsheet section follows. The discussion is about (1) how each section should be set up, and (2) the logic of the formulas in the section's cells.

CHANGING CELLS Section

Your spreadsheet should have the changing cells shown in Figure 10-7.

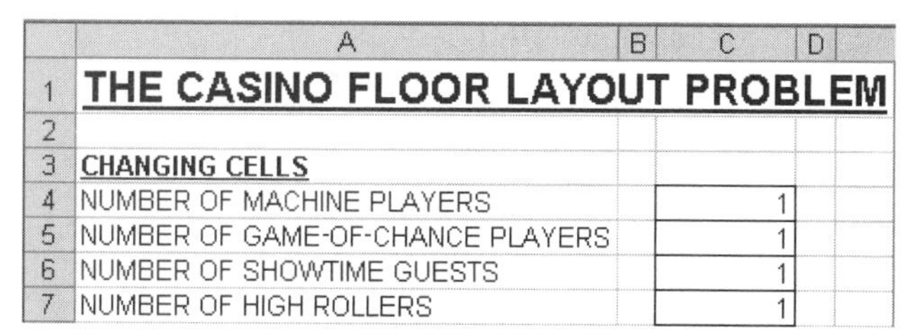

	A	B	C	D
1	THE CASINO FLOOR LAYOUT PROBLEM			
2				
3	CHANGING CELLS			
4	NUMBER OF MACHINE PLAYERS		1	
5	NUMBER OF GAME-OF-CHANCE PLAYERS		1	
6	NUMBER OF SHOWTIME GUESTS		1	
7	NUMBER OF HIGH ROLLERS		1	

Figure 10-7 CHANGING CELLS section

You are asking the Solver model to compute the number of each kind of casino guest that management should want. Start with a *1* in each cell. The Solver will change each *1* as it computes the answer. Do not allow the Solver to recommend a fraction of a guest.

CONSTANTS Section

Your spreadsheet should have the constants shown in Figures 10-8 and 10-9. An explanation of the line items follows each figure.

Case 10

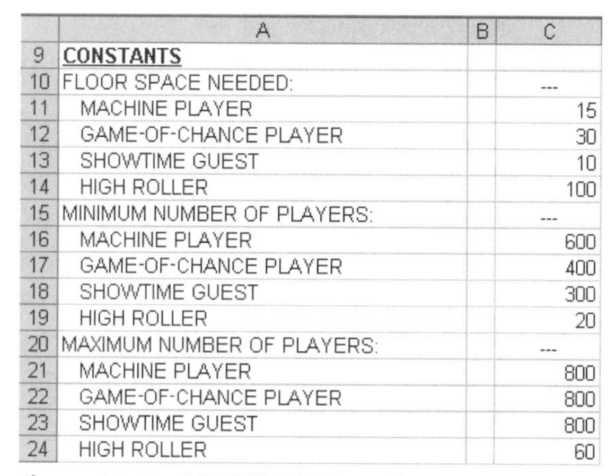

	A	B	C
9	CONSTANTS		
10	FLOOR SPACE NEEDED:		---
11	MACHINE PLAYER		15
12	GAME-OF-CHANCE PLAYER		30
13	SHOWTIME GUEST		10
14	HIGH ROLLER		100
15	MINIMUM NUMBER OF PLAYERS:		---
16	MACHINE PLAYER		600
17	GAME-OF-CHANCE PLAYER		400
18	SHOWTIME GUEST		300
19	HIGH ROLLER		20
20	MAXIMUM NUMBER OF PLAYERS:		---
21	MACHINE PLAYER		800
22	GAME-OF-CHANCE PLAYER		800
23	SHOWTIME GUEST		800
24	HIGH ROLLER		60

Figure 10-8 CONSTANTS section

- FLOOR SPACE NEEDED: Each guest requires a certain amount of the available floor space.
- MINIMUM AND MAXIMUM NUMBER OF PLAYERS: There are minimums and maximums for each kind of casino guest.

Constants are continued in Figure 10-9

	A	B	C
25	BASE REVENUE PER DAY:		---
26	MACHINE PLAYER		200
27	GAME-OF-CHANCE PLAYER		300
28	SHOWTIME GUEST		100
29	HIGH ROLLER		5000
30	FOOD REVENUE PER DAY:		---
31	MACHINE PLAYER		25
32	GAME-OF-CHANCE PLAYER		100
33	SHOWTIME GUEST		125
34	HIGH ROLLER		0
35	DIRECT EXPENSES PER DAY		---
36	MACHINE PLAYER		1
37	GAME-OF-CHANCE PLAYER		40
38	SHOWTIME GUEST		10
39	HIGH ROLLER		500
40	TOTAL ROOM SQUARE FOOTAGE		50000
41	STAGE SQUARE FOOTAGE		5000
42	DINING AREA SQUARE FOOTAGE		5000
43	FIRE MARSHAL LIMIT		1850
44	HOTEL OCCUPANCY		1080

Figure 10-9 CONSTANTS section—continued

- BASE REVENUE PER DAY: Each guest can be expected to generate a certain amount of gambling and/or ticket revenue each day. Average amounts for each type of guest are shown.
- FOOD REVENUE PER DAY: Each guest can be expected to generate a certain amount of food revenue each day. Average amounts for each type of guest are shown.
- DIRECT EXPENSES PER DAY: Each guest can be expected to create a certain level of extra expense each day. Average amounts for each type of guest are shown.
- TOTAL ROOM SQUARE FOOTAGE: The entire ground-floor space covers 50,000 square feet.
- STAGE SQUARE FOOTAGE: On one end of the room, 1500 feet are set aside for a stage and related dining.
- DINING AREA SQUARE FOOTAGE: On the other end of the room is a 5000-square-foot dining area where gamblers can eat.
- FIRE MARSHAL LIMIT: No more than 1850 people can crowd into the casino at any one time.
- HOTEL OCCUPANCY: The hotel can hold 1850 guests. The casino must be large enough to hold at least its hotel occupants, in case they should all want to use the casino at the same time.

CALCULATIONS Section

Your spreadsheet should calculate the amounts shown in Figure 10-10. They would be used in the INCOME STATEMENT section and/or in the CONSTRAINTS section. Calculations are based on values in the CHANGING CELLS and/or CONSTANTS sections, and/or on other calculated values. An explanation of the line items follows the figure.

	A	B	C
46	**CALCULATIONS**		
47	SPACE AVAILABLE FOR GUESTS		
48	FLOOR SPACE USED:		---
49	MACHINE PLAYERS		
50	GAME-OF-CHANCE PLAYERS		
51	SHOWTIME GUESTS		
52	HIGH ROLLERS		
53	TOTAL FLOOR SPACE USED FOR GUESTS		
54	BASE REVENUE EARNED PER DAY:		---
55	MACHINE PLAYERS		
56	GAME-OF-CHANCE PLAYERS		
57	SHOWTIME GUESTS		
58	HIGH ROLLERS		
59	FOOD REVENUE EARNED PER DAY:		---
60	MACHINE PLAYERS		
61	GAME-OF-CHANCE PLAYERS		
62	SHOWTIME GUESTS		
63	HIGH ROLLERS		
64	DIRECT EXPENSES INCURRED PER DAY		---
65	MACHINE PLAYERS		
66	GAME-OF-CHANCE PLAYERS		
67	SHOWTIME GUESTS		
68	HIGH ROLLERS		
69	REVENUE ADJUSTMENTS:		---
70	MACHINE PLAYERS		
71	GAME-OF-CHANCE PLAYERS		
72	SHOWTIME GUESTS		
73	HIGH ROLLERS		
74	TOTAL NUMBER OF GUESTS		

Figure 10-10 CALCULATIONS section

- SPACE AVAILABLE FOR GUESTS: The space available is the total area less room for dining and the stage. All the casino guests must fit into the space available.
- FLOOR SPACE USED: Calculate how much space is used by all machine players, by all game-of-chance players, and so forth.
- TOTAL FLOOR SPACE USED FOR GUESTS: Calculate the total of floor space used by all types of guests.
- BASE REVENUE EARNED PER DAY: Compute base revenue earned by all machine players, by all game-of-chance players, and so forth.
- FOOD REVENUE EARNED PER DAY: Compute food revenue earned by all machine players, by all game-of-chance players, and so forth.
- DIRECT EXPENSES INCURRED PER DAY: Compute direct expenses incurred by all machine players, by all game-of-chance players, and so forth.
- REVENUE ADJUSTMENTS: For each kind of guest, compute revenue adjustments following the rules for changing the propensity to spend.
- TOTAL NUMBER OF GUESTS: Compute the total of all casino guests in the hotel.

INCOME STATEMENT Section

Compute total revenues and expenses for a day for each kind of guest, as shown in Figure 10-11. Line items are discussed after the figure.

	A	B	C
76	**INCOME STATEMENT**		
77	TOTAL BASE REVENUE		
78	FOOD REVENUE		
79	REVENUE ADJUSTMENTS		
80	TOTAL REVENUE		
81	DIRECT EXPENSES		
82	INCOME EARNED		

Figure 10-11 INCOME STATEMENT section

- TOTAL BASE REVENUE: This is the total of base revenue for all kinds of guests.
- FOOD REVENUE: This is the total of food revenue for all kinds of guests.
- REVENUE ADJUSTMENTS: This is the total of revenue adjustments for all kinds of guests.
- TOTAL REVENUE: This is the total of all kinds of revenue.
- DIRECT EXPENSES: This is the total of direct expenses for all kinds of guests.
- INCOME EARNED: This is total revenue less direct expenses.

CONSTRAINTS and Running the Solver

Determine the constraints. Enter the Base Case decision constraints, using the Solver. Run the Solver. Check the logic of the default values for the Solver's setup. Make an Answer Report when the Solver says that a solution has been found that satisfies the constraints.

When you are done, print the entire workbook, including the Answer Report. Save the Base Case Solver spreadsheet file (File—Save). Then, to prepare for the Extension Case, go to your Data Files and select the spreadsheet skeleton **FLOOR2.XLS**.

Assignment 1B: Creating the Spreadsheet—Extension Case

As previously noted, some members of management think that less space should be allocated to the large gambling dining area, and they also think there should be no upper limit on the number of high rollers or the number of game-of-chance gamblers.

You should modify the Extension Case spreadsheet (**FLOOR2.xls**) to accommodate this alternative. Run the Solver. Ask for an Answer Report when the Solver says that a solution has been found that satisfies the constraints. When you are done, print the entire workbook, including the Solver Answer Report. Save the workbook. Close the file and Exit Excel.

Assignment 2 Using the Spreadsheet for Decision Support

You have built the Base Case and the Extension Case models because you want to know the guest mix in each scenario and which scenario yields the highest income earned. You will now complete the case by (1) using the Answer Reports to gather the data needed to decide which floor setup to use; (2) documenting your recommendation in a memorandum; and (3), if your instructor specifies, an oral presentation.

Assignment 2A: Using the Spreadsheet to Gather Data

You have printed the spreadsheet and Answer Report sheet for each scenario. You can see how many of each kind of casino guest to attract for each case. Note the data for each case on a separate sheet of paper for inclusion in a table in your memorandum. The form of your table is shown in Figure 10-12.

	Base Case (large dining area)	Extension Case (smaller area)
Machine players		
Game-of-chance players		
Showtime guests		
High rollers		
Total number of guests		
Income earned		

Figure 10-12 Form for table in memorandum

Assignment 2B: Documenting Your Recommendation in a Memorandum

Write a brief memorandum to HEC management about the results. Observe the following requirements:

- Your memorandum should have a proper heading (DATE / TO / FROM / SUBJECT). You may wish to use a Word memo template (**File**—click **New**, click **On my computer** in the Templates section, click the **Memos** tab, choose **Contemporary Memo**, then click **OK**).
- Briefly outline the situation and the recommended decisions.
- A casino floor that has a lot of high rollers and machine players would have a very different look than a floor with a lot of showtime guests and few machine players, for example. Tell management how you think the casino would look to an outside observer in each case.
- State which floor configuration gives the higher income earned.
- Management would want to go with the configuration that yields the highest income earned, unless the alternatives result in about the same income earned. In that case, management would want to go with the alternative that has the more wholesome, civilized look. Tell management which alternative they should adopt.

Support your recommendation graphically by making a summary table in Word after the first paragraph, such as that shown in Figure 10-12.

Enter a table into Word, using the following procedure:

1. Select the **Table** menu option, point to **Insert**, then click **Table**.
2. Enter the number of rows and columns.
3. Select **AutoFormat** and choose **Table Grid 1**.
4. Select **OK**, and then select **OK** again.

Case 10

Assignment 3 Giving an Oral Presentation

Assume that HEC management members are impressed by your analysis. They ask that you give a presentation explaining your recommendation to a wider group of company management and employees. Prepare to explain your analysis and recommendation to the group in 10 minutes or fewer. Use visual aids or handouts that you think are appropriate. Tutorial E has guidance on how to prepare and give an oral presentation.

DELIVERABLES

Assemble the following deliverables for your instructor:

1. Printout of your memorandum
2. Spreadsheet printouts
3. Disk or CD, which should have your Word memorandum file and your Excel spreadsheet files

Staple the printouts together, with the memorandum on top. If there are other .XLS files on your disk or CD, write your instructor a note, stating the names of your Base Case and Extension Case .xls files.

Decision Support Cases Using Basic Excel Functionality

Master Blaster Division Operating Budget

Decision Support Using Excel

Preview

World Wide Games Corporation makes and sells action games and toys. Two years ago, the Master Blaster Division introduced a paintball gun, and the product has sold well. In this case, you will use Excel to make an operating budget for the coming year that will forecast quarterly net income and cash on hand. The budget will be based on three different ways of forecasting units sold, sales forecasts, and three possible states of the economy.

Preparation

- Review spreadsheet concepts discussed in class and/or in your textbook. This case requires an understanding of these Excel functions: IF(), SUM(), AVERAGE(), and TREND().This case also requires you to understand these Excel capabilities: Analysis Toolpak Add-In, Moving Average Tool, Array functions, Charting data, and Importing Data.
- Complete any exercises that your instructor assigns.
- Complete any parts of Tutorials C and D that your instructor assigns, or refer to them as necessary.
- Review file-saving procedures for Windows programs. These are discussed in Tutorial C.
- Refer to Tutorial E as necessary.

BACKGROUND

World Wide Games Corporation (WWG) makes and sells a variety of games for "kids of all ages." The corporation's promotions claim, "When you think of *action* and *games* you think of World Wide Games."

Two years ago, WWG's Master Blaster Division debuted a new and improved paintball gun. The gun can lob a pouch of paint much farther than competing guns, and the accuracy is incredible. Sales have been brisk. Operating data for the product's first four quarters are summarized in Figure 11-1. An explanation of line items follows the figure.

	Q1	Q2	Q3	Q4
Units Sold	1,000,000	980,000	970,000	1,200,000
Unit Selling Price	$100.00	$99.00	$99.50	$101.00
Unit Variable Cost	$70.00	$71.00	$72.00	$73.00
Fixed Cost	$10,000,000	$10,000,000	$10,000,000	$10,000,000
Tax Rate	.30	.30	.30	.30
Assets Invested	$50,000,000	$50,000,000	$50,000,000	$50,000,000

Figure 11-1 Operating data for first four quarters

Case 11

- Units Sold is the number of paintball guns made and sold in each quarter's three months.
- Unit Selling Price is the average selling price of a gun in the quarter.
- Unit Variable Cost is the cost of labor and materials expended when a gun is made.
- Fixed Cost is the divisional costs incurred, regardless of how many units are made and sold.
- Tax Rate is the corporation's overall tax rate (state and federal) on net income.
- Assets Invested is the value of the division's property and equipment devoted to making and selling paintball guns. This is the amount of total assets on the divisional balance sheet at the end of a quarter.

The operating data for the product's second four quarters are summarized in Figure 11-2

	Q5	Q6	Q7	Q8
Units Sold	1,100,000	990,000	1,100,000	1,300,000
Unit Selling Price	$100.50	$101.20	$102.00	$101.50
Unit Variable Cost	$74.00	$75.00	$76.00	$77.00
Fixed Cost	$11,000,000	$11,000,000	$11,000,000	$11,000,000
Tax Rate	.33	.33	.33	.33
Assets Invested	$60,000,000	$60,000,000	$60,000,000	$60,000,000

Figure 11-2 Operating data for second four quarters

As you can see, more assets were devoted to the business in the second year when it became clear that the business was going to be a good one.

You can assume that it is the beginning of the third year, and division management wants to see a forecast of net income and cash generated by the business for the coming four quarters. The first step in making an operating forecast is forecasting units sold. Company management wants third-year units sold to be forecasted in three ways:

1. A simple average of the first eight quarters' units sold
2. A moving average of the first eight quarters' units sold
3. A linear regression of the first eight quarters' units sold

At this point, a digression is in order to discuss the differences in the three ways of forecasting.

Methods of Forecasting

An **average** of a series of values is its typical, or "central," value. Most people compute a **mean** when they want an average. You already know how to compute a mean. For example, in this case, add sales for all eight months and divide the total by eight. Excel's AVERAGE() function will do that for you. Note that doing this assumes that each of the eight values is weighted equally—each is as important as the other.

Moving average and linear regression are estimation techniques that try to smooth out fluctuations in a series of data values. These methods are explained next.

Calculating a Moving Average

The **moving average** technique is most easily explained by an example. Suppose that you have sales data for five prior months. You want to make a three-month forecast based on the five prior months' data. The first of the three months' forecast is obtained by averaging the revenue of prior months 1, 2, and 3. The second of the three months' forecast is obtained by averaging the revenue of prior months 2, 3, and 4. The third of the three months' forecast is obtained by averaging the revenue of prior months 3, 4, and 5. Figure 11-3 shows some example data.

	A	B	C	D	E	F
1	MOVING AVERAGE EXAMPLE					
2		1	2	3	4	5
3	REVENUE	1100	1050	980	1200	1150
4						
5		6	7	8		
6	FORECASTED REVENUE					

Figure 11-3 Moving Average example

Looking at the data in Figure 11-3, you can see from the five months of data that the trend is generally up (the last two months are higher than the first three), but the data values do fluctuate. A moving average "smooths" out the fluctuations. Let's insert some numbers: the value for month 6 would be based on months 1, 2, and 3: (1100 + 1050 + 980) / 3 = 1043. The value for month 7 would be based on months 2, 3, and 4: (1050 + 980 + 1200) / 3 = 1077. The value for month 8 would be based on months 3, 4, and 5: (980 + 1200 + 1150) / 3 = 1110.

Notice that you could plot the three values: 1043, 1077, and 1110. Thus, the average "moves." The plot of the data points would show less variability (scatter) than a plot of the raw data, which is why a moving average is said to "smooth" out data. The trend of the plot (the "movement," as it were) depends, in part, on how many values are used to compute each average—here, three values are used to compute each value. The mean is a single value based on all data values, but moving average is not. Notice also that a moving average uses more recent values and ignores some older values. Here, for example, the first two month's values are not used to compute the last value in the moving average (1110). This contrasts with computing a mean, which assumes all values in the series matter and are equally weighted.

Computing a moving average in Excel probably will require three steps:

1. You must first install the Analysis ToolPak, which has the moving average code. To install, take this path: **Tools—Add-Ins—Analysis ToolPak**. (You only need to install the Analysis Tool Pak once.)
2. To actually use the moving average tool, take this path: **Tools—Data Analysis—Moving Average —OK**. Then specify three values: (a) the range of the base data used in the forecast; (b) the range of cells that will hold the output; and (c) the size of the data width interval (in the example, the interval was 3—three months' values were used to compute one month's value). Click **OK** to get the forecasted values in the output range.
3. The outputs will include some *NA* values in the beginning of the range—the forecasted values will be at the end of the range. You should use **Edit—Copy** to copy the forecasted values (without the *NA*'s), then go to **Edit—Paste Special—Values** to copy those values to the intended range in the spreadsheet.

These steps for the example are shown next. Selecting **Tools—Data Analysis—Moving Average—OK** brings up the window shown in Figure 11-4.

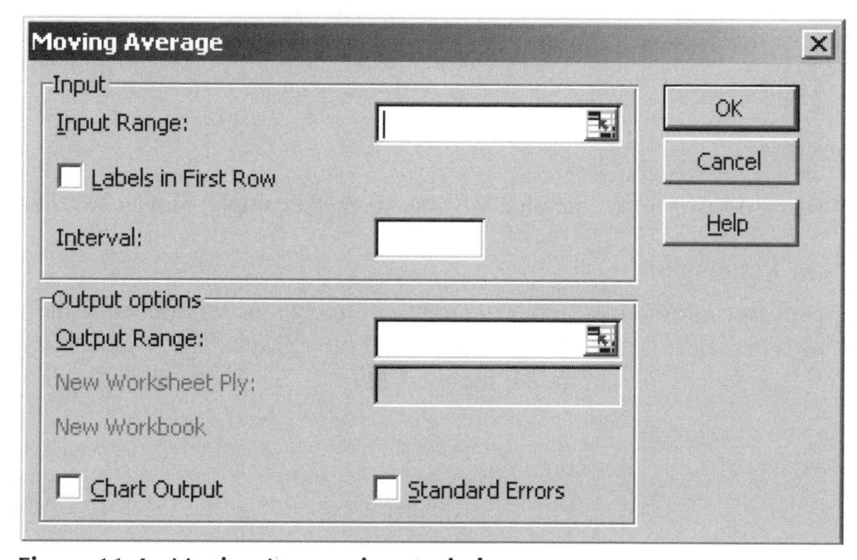

Figure 11-4 Moving Average input window

In the example, the Input Range is B3 to F3. We want a three-month moving average, so the value to put in the Interval box is 3. The Output Range is B6 to F6. The entries are shown in Figure 11-5. (Excel will put dollar signs in the ranges—do not worry about this.)

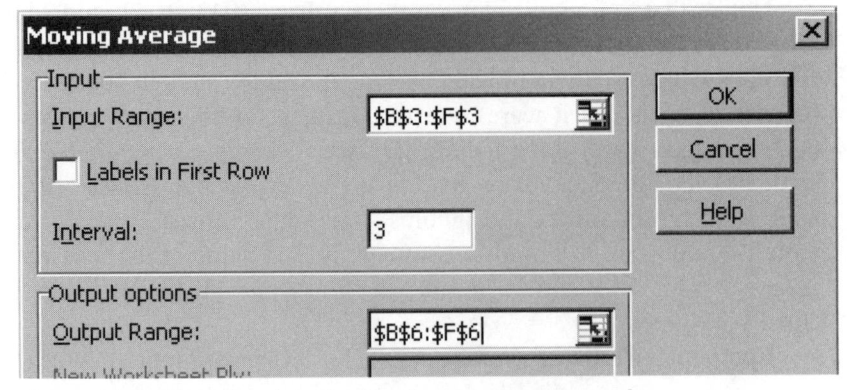

Figure 11-5 Entering data for Moving Average example

Clicking OK causes the output data to appear as shown in Figure 11-6.

	A	B	C	D	E	F
1	MOVING AVERAGE EXAMPLE					
2		1	2	3	4	5
3	REVENUE	1100	1050	980	1200	1150
4						
5		6	7	8		
6	FORECASTED REVENUE	#N/A	#N/A	1043	1077	1110

Figure 11-6 Output data for Moving Average example

There will always be some *NA*s in the output. (The number of these will be the size of the interval less 1.) This is shown in Figure 11-6. Apparently, there is no way to avoid the *NA* values—specifying B6..D6 as the output range will not overcome the problem. You can cut and paste the real data to the proper cells, as shown in Figure 11-7.

	A	B	C	D	E	F
1	MOVING AVERAGE EXAMPLE					
2		1	2	3	4	5
3	REVENUE	1100	1050	980	1200	1150
4						
5		6	7	8		
6	FORECASTED REVENUE	1043	1077	1110		

Figure 11-7 Output data for Moving Average example, moved to correct positions

Linear Regression

Linear regression is another smoothing technique. Again, imagine the same five months of prior sales data, with the data plotted on a graph. The Y-axis values are dollars, and the X-axis values are the month numbers (1, 2, ... 5). The scatter plot is shown in Figure 11-8.

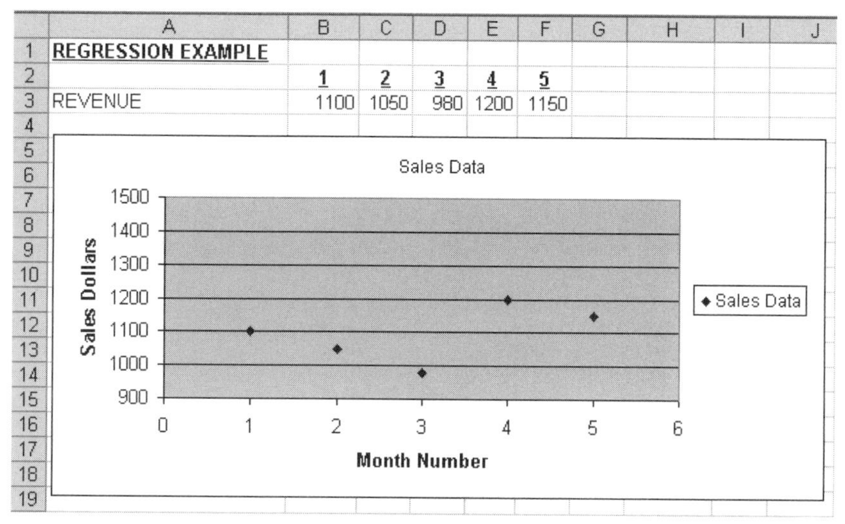

Figure 11-8 Scatter plot of example sales data

Notice that the data values fluctuate. Linear regression finds the straight line through the data that "fits" in the five data points the best. The straight line is, thus, a way of smoothing out fluctuations in the values. Figure 11-9 shows the linear regression line through the sample data that Excel would compute.

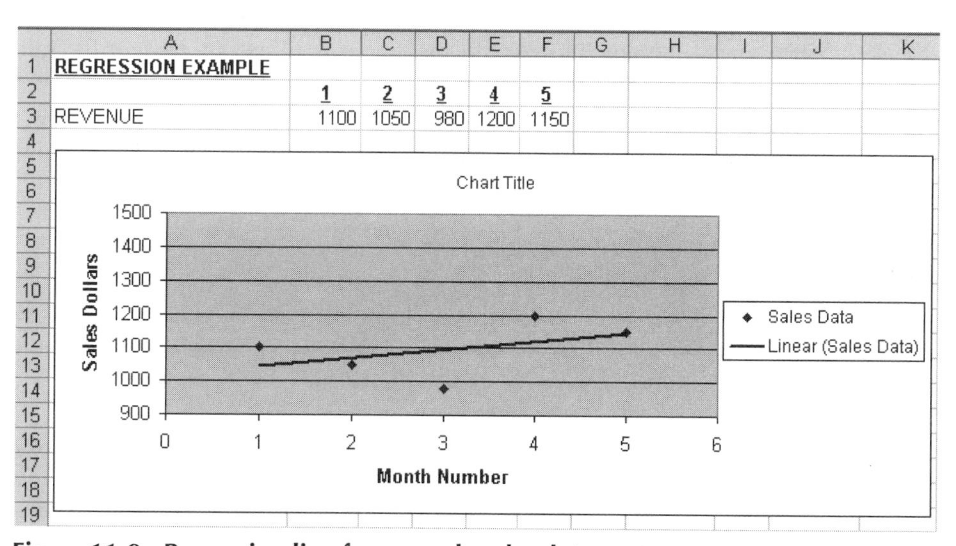

Figure 11-9 Regression line for example sales data

The formula for a straight line is: $Y = mX + b$, where m is the slope of the line and b is the Y-intercept of the line. The Y values depend on the X values and on the slope and the Y-intercept. Figure 11-10 shows the formula for the regression line that Excel would compute.

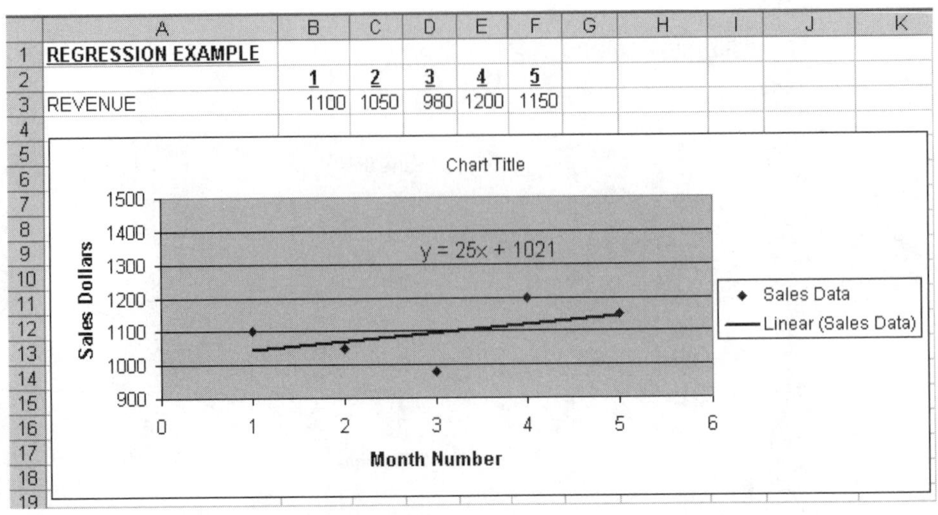

Figure 11-10 Regression line for example sales data, with formula

The straight line can be used to estimate future values. Here, for Month 6, estimated sales would be: 25 * 6 + 1021 = 1171. Notice that the positive slope (here, +25) indicates that sales are assumed to be increasing in all future periods. If the slope were negative, the line would be "tilted" downward from left to right. Forecasted sales would be decreasing.

How closely does the line "fit" the data points? The accuracy of the fit is indicated by the "R-squared" value, which can range from 0 to 1. A value of 0 means that the line is a complete guess. A value of 1 would mean that the line is a perfect fit—all data points would be *on* the line. An R-squared near zero means the line is not a good predictor. An R-squared close to 1 means the line is probably a good predictor. Excel can show the "R-squared" value of the line. Figure 11-11 shows the R-squared value for the example line.

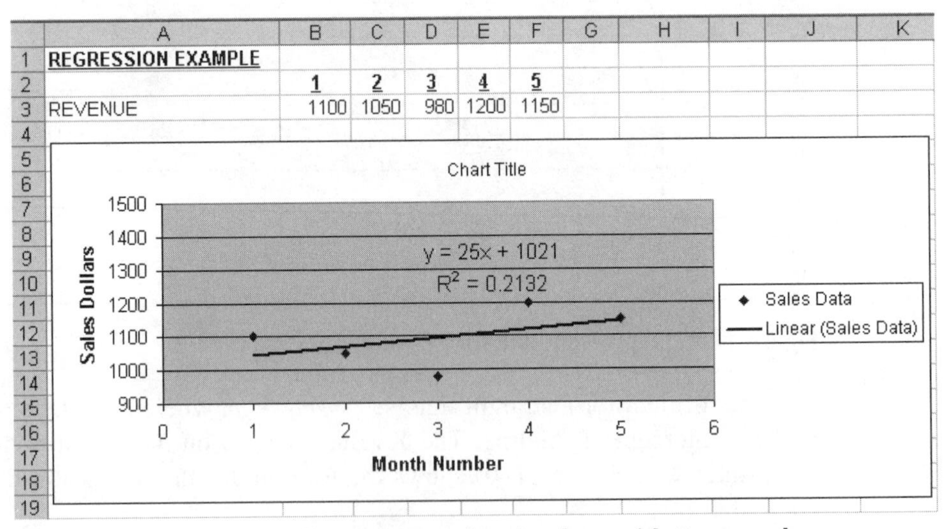

Figure 11-11 Regression line for example sales data, with R-squared

In the example, an R-squared value of .2132 suggests that the line does not fit the data very well.

Linear regression finds the straight line through the data that "fits" in the available five data points the best. You could then obtain the predicted three months' data in one of two ways:

1. The line in the graph would then be extended for the next three months, and the dollar values read off of the graph for months 6, 7, and 8, as was shown in the previous figure.
2. A formula for the straight line could be obtained without making a scatter plot and then would be used to compute the predicted values. The formula for a straight line is: $Y = mX + b$, where m is the slope of the line and b is the Y-intercept of the line. Here, X values would be 6, 7, and 8 (for the months). Y values would be the forecasted sales values; m and b can be estimated by looking at the graph, or computed precisely. Excel's LINEST() function outputs the m and b coefficients associated with a set of base data. Forecasted Y values can then be computed in Excel, using the m and b coefficients. More straightforwardly, Excel's TREND() function can be used to compute the forecasted Y values directly. Excel's RSQ() function can be used to compute the R-squared value directly.

TREND() is nicely explained in Excel's Help system. To see Help, click the *Fx* function icon in the formula bar. Type in the name of the function in the window, and click the GO button. You will get a window that gives the function's syntax. You will also get a link to *Help on this function*, which you should click for an extended explanation, including examples, which are usually very helpful.

The prior example is redone here, using =TREND(). The =TREND() function syntax is as follows:

=TREND(known Y values, known X values, new X values)

Here, B3..F3 are the known Y values, B2..F2 are the known X values. The new X values are 6, 7, and 8 in cells B5..D5. The function =TREND() is a so-called "array" function, which means it uses a range of values for its outputs. Array functions are entered somewhat differently. The first step is to enter the formula as if for a regular function, in the first target output cell, in this case, B6. This is shown in Figure 11-12:

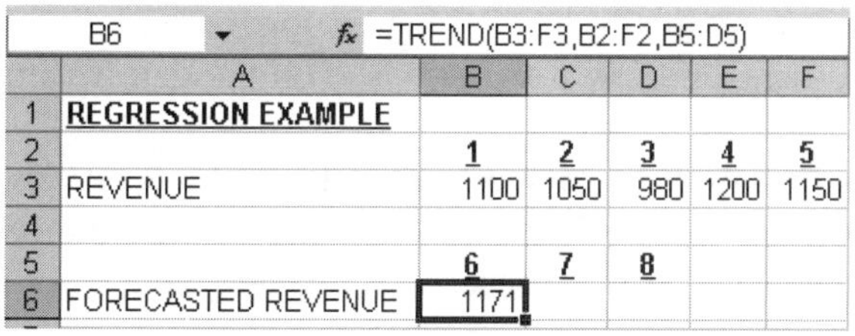

B6 fx =TREND(B3:F3,B2:F2,B5:D5)

	A	B	C	D	E	F
1	REGRESSION EXAMPLE					
2		1	2	3	4	5
3	REVENUE	1100	1050	980	1200	1150
4						
5		6	7	8		
6	FORECASTED REVENUE	1171				

Figure 11-12 Entering the =TREND() formula

The next step is to highlight the output range, by clicking and dragging the cursor—here, B6 to D6. Then, press the F2 key. This causes the screen to change, as shown in Figure 11-13.

28 TREND ▾ X ✓ fx =TREND(B3:F3,B2:F2,B5:D5)

	A	B	C	D	E	F
1	REGRESSION EXAMPLE					
2		1	2	3	4	5
3	REVENUE	1100	1050	980	1200	1150
4						
5		6	7	8		
6	FORECASTED REVENUE	=TREND(B3:F3,B2:F2,B5:D5)				

Figure 11-13 Entering the =TREND() formula—F2 pressed

The next step is to press Ctrl, Shift, and Enter at the same time. This causes the output to be sent to the three highlighted cells as shown in Figure 11-14:

	A	B	C	D	E	F
1	REGRESSION EXAMPLE					
2		1	2	3	4	5
3	REVENUE	1100	1050	980	1200	1150
4						
5		6	7	8		
6	FORECASTED REVENUE	1171	1196	1221		

Figure 11-14 Entering the =TREND() formula—Ctrl+Shift+Enter pressed

The R-squared value could be computed using the RSQ() function. The syntax of that is: =RSQ(known *Y* values, known *X* values). Here, =RSQ(B3..F3, B2..F2) would yield a value of .2132.

You should notice that the moving average and linear regression forecasts give different results. Moving average gave these results: 1043, 1077, and 1110; linear regression gave these results: 1171, 1196, 1221. Moving average smooths out fluctuations but does not eliminate fluctuations. The function =TREND() sees a positively sloping line through the fluctuating data values—the line moves from Phoenix to Boston, to use a geographical metaphor. All projected values will be along the line, increasing in value for each *X*.

What happens if forecasting methods yield different forecasted values? How does the user know which values are the "right" ones? Here is some guidance on how to answer that question.

One reality check is to compute the R-squared for the data values. A poor R-squared indicates very scattered data, and any forecast based on the data will be suspicious (not just the linear regression).

The user must apply judgment to see how well the logic of the underlying situation matches with the forecasted values. The mean assumes that all the data values are equally valid for the forecast. However, if more recent data values seem to be better predictors than older data values, then a moving average is probably better. A moving average uses a certain number of data values to predict the next value in the series—but does the interval seem big enough? A positive regression-line slope suggests that all forecasted values will be increasing. Does this seem a reasonable likelihood? A negative slope indicates constantly decreasing forecasted values. Is this reasonable? How can the user answer these questions? Probably not by merely looking at the statistics alone. The user would have to investigate the logic of the situation by other means—talking to people, examining the results of questionnaires, or by other forms of research.

Let us return now to the WWG situation, the Master Blaster Division's operating budget. Forecasted units sold and sales price can be used to compute forecasted sales. Cost of goods sold, interest expense, and taxes can be estimated and used to compute net income after taxes.

The Operating Budget

Management assumes that sales will be collected in cash in the quarter in which the income is earned (i.e., customers will pay their bills quickly). Net income after taxes can then be estimated. Cash on Hand equals Beginning Cash plus Net Income. Management is expected to generate a certain amount of cash in each quarter. If the division falls short of the required cash, the company will "lend" enough cash to the division to get back to the required minimum level—it is as if the company acts as a bank for each division. Interest Expense is charged as a cost by the company for any such borrowings, however—just as a bank would do! Company management keeps track of such internal debt—a division is expected to pay its way over time, of course.

A key measure of divisional management performance is its Return on Investment (ROI), which is the division's net income divided by the value of the assets invested in the business. Sometimes this ratio is called Return on Assets (ROA). If the ratio is not high enough, company management will need to think about investing in other businesses.

For company (and division) management, one of the major reasons for creating an operating budget is to see what returns appear likely in the future. If returns look good, they will invest more in the business. If returns look poor, they will decide whether the business can be fixed, or if it needs to be abandoned. Thus, developing an operating budget for the next year is just part of good management practice.

Assignment 1 Creating a Spreadsheet for Decision Support

In this assignment, you will produce a spreadsheet that models the business problem—the need for a forecast of the business in the next four quarters. Then, in Assignment 2, you will write a memorandum to management that explains your findings and gives recommended actions. In addition, in Assignment 3, you will be asked to prepare an oral presentation of your analysis and recommendation.

Now you will create the forecast in Excel. You will be given some hints on how each section should be set up before entering cell formulas. Your spreadsheet should have the following sections.

- **HISTORICAL DATA (CONSTANTS)**
- **INPUTS**
- **SUMMARY OF KEY RESULTS**
- **CALCULATIONS**
- **INCOME STATEMENT AND CASH FLOW STATEMENT**
- **DEBT OWED**

A discussion of each section follows. *The spreadsheet skeleton is available to you, so you need not type in the skeleton if you do not want to do so.* To access the spreadsheet skeleton, go to your Data files. Select Case 11, then select **BUDGET.xls.**

HISTORICAL DATA (CONSTANTS) Section

Your spreadsheet should have the historical data previously described, as Constants. To import that data, you should make an input file in .txt format. This can be done using Notepad (Start—All Programs—Accessories—Notepad). This is shown in Figure 11-15.

```
history.txt - Notepad
File  Edit  Format  Help
1000000, 980000, 970000, 1200000, 1100000, 990000, 1100000, 1300000
100.00, 99.0, 99.5, 101.0, 100.5, 101.2, 102.0, 101.5
70.00, 71.0, 72.0, 73.0, 74.0, 75.0, 76.0, 77.0
10000000, 10000000, 10000000, 10000000, 11000000, 11000000, 11000000, 11000000
.30, .30, .30, .30, .33, .33, .33, .33
50000000, 50000000, 50000000, 50000000, 60000000, 60000000, 60000000, 60000000
```

Figure 11-15 Historical data in Notepad file

This data should then be imported into Excel. The section initially looks like Figure 11-16:

	A	B	C	D	E	F	G	H	I	J
1	**MASTER BLASTER DIVISION OPERATING BUDGET**									
2	HISTORICAL DATA				PRIOR YEARS' QUARTERS					
3			1	2	3	4	5	6	7	8
4	UNITS SOLD	NA								
5	UNIT SELLING PRICE	NA								
6	UNIT VARIABLE COST	NA								
7	FIXED COSTS	NA								
8	TAX RATE	NA								
9	ASSETS INVESTED	NA								

Figure 11-16 HISTORICAL DATA section before importing data

To import data from the text file, follow this path:

1. Choose **Data—Import External Data—Import Data.**
2. Select the file in the **Look in** window.
3. Specify that the file is a comma delimited file, **General Format.**
4. Specify the data range's upper left cell, in this case **C4**. Once imported, the historical data section would look like Figure 11-17.

	A	B	C	D	E	F	G
1	**MASTER BLASTER DIVISION OPERATING BUDGET**						
2	HISTORICAL DATA				PRIOR YEARS' QUARTERS		
3			1	2	3	4	5
4	UNITS SOLD	NA	1000000	980000	970000	1200000	1100000
5	UNIT SELLING PRICE	NA	100	99	99.5	101	100.5
6	UNIT VARIABLE COST	NA	70	71	72	73	74
7	FIXED COSTS	NA	10000000	10000000	10000000	10000000	11000000
8	TAX RATE	NA	0.3	0.3	0.3	0.3	0.33
9	ASSETS INVESTED	NA	50000000	50000000	50000000	50000000	60000000

Figure 11-17 HISTORICAL DATA section after importing data

In the figure, only the first five quarters are shown. Your spreadsheet would have data for eight quarters, of course.

INPUTS Section

Your spreadsheet should have the inputs shown in Figure 11-18. Your instructor may tell you to apply Conditional Formatting to the input cells, so that out-of-bounds values are highlighted in some way (for example, the entry shown in red and/or in boldface type). If so, your instructor may provide a tutorial on Conditional Formatting. Or, your instructor may ask you to refer to Excel Help. An explanation of the line items follows the figure.

	A	B	C	D	E	F
11	INPUTS			CURRENT YEAR QUARTERS		
12			9	10	11	12
13	UNITS SOLD FORECAST METHOD (1 = SIMPLE AVG; 2 = MOVING AVG; 3 = REGRESSION)	NA		NA	NA	NA
14	ECONOMY (G = GOOD; O = OK; B = BAD)	NA				

Figure 11-18 INPUTS section

- UNITS SOLD FORECAST METHOD: You will indicate the way to forecast units sold. Enter a *1* for a simple average, a *2* for a moving average, and a *3* for a regression-based forecast. The entry applies to all four forecast periods, months 9 through 12.
- ECONOMY: You will indicate the expected state of the economy in each quarter. A ***G*** indicates a Good, healthy economy for sellers of games and toys. An ***O*** indicates an OK economy. A ***B*** indicates a Bad, or poor, economy for such companies.

Case 11

SUMMARY OF KEY RESULTS Section

Your spreadsheet should show the results shown in Figure 11-19.

	A	B	C	D	E	F
16	SUMMARY OF KEY RESULTS		9	10	11	12
17	NET INCOME AFTER TAXES	NA				
18	END-OF-THE-QTR CASH ON HAND	NA				
19	END-OF-THE-QTR DEBT OWED	NA				
20	RETURN ON ASSETS IN QUARTER	NA				

Figure 11-19 SUMMARY OF KEY RESULTS section

In this section, values are echoed from other areas of the spreadsheet. For each quarter, you should show the following:

- NET INCOME AFTER TAXES, from the Income Statement
- END-OF-THE-QTR CASH ON HAND, from the Cash Flow Statement
- END-OF-THE-QTR DEBT OWED, from the Debt Owed Section
- RETURN ON ASSETS IN QUARTER, a calculation

Your instructor may tell you to chart (graph) some key outputs. If so, you will need a place in your spreadsheet to put the data that will be charted. A good place would be the area to the right of the SUMMARY OF KEY RESULTS section. For example, your instructor may tell you to chart Quarter 12's Returns on Investments for each Economy and Units Sold forecasting method. You could set up an area like that shown in Figure 11-20.

	CHARTING: QTR 12 RETURN ON ASSETS			
	ECONOMY	SIMPLE AVG	MOVING AVG	REGRESSION
	GOOD			
	OK			
	BAD			

Figure 11-20 Charting data area

You would enter the input combinations (e.g., the return with a Good economy and units based on a simple average —*1*). You would note the Quarter 12 Return on Investment value, and manually enter it into the appropriate charting area cell. You would do this for all nine input combinations. When finished, you will have a data area that looks like that shown in Figure 11-21.

	CHARTING: QTR 12 RETURN ON ASSETS			
	ECONOMY	SIMPLE AVG	MOVING AVG	REGRESSION
	GOOD	19.05	20.66	23.83
	OK	5.68	6.5	8.58
	BAD	-13.25	-13.27	-11.96

Figure 11-21 Completed Charting data area

These values could then serve as the basis of a chart. (Here, values are illustrative.)

CALCULATIONS Section

You should calculate various intermediate results, which are then used in other calculations, or in the net income and cash flow statement that follows. Calculations are based on historical data and your inputs. When called for, use absolute addressing properly. Calculate the values shown in Figure 11-22. An explanation of the line items follows the figure.

	A	B	C	D	E	F
22	CALCULATIONS		9	10	11	12
23	ASSETS INVESTED	NA				
24	CASH NEEDED TO START QUARTER	NA				
25	UNITS SOLD - MOVING AVG	NA				
26	UNITS SOLD - SIMPLE AVG	NA				
27	UNITS SOLD - REGRESSION	NA				
28	SELLING PRICE	$101.50				
29	UNIT VARIABLE COST	$77.00				
30	INTEREST RATE	NA				
31	RETURN ON ASSETS INVESTED	NA				

Figure 11-22 CALCULATIONS section

- ASSETS INVESTED: This is the value of the assets invested in the division at the beginning of the quarter. In Quarter 9, assume that this equals the assets at the end of Quarter 8. For succeeding quarters, assume that assets invested equal the prior quarter's assets plus the prior quarter's net income retained in the business. Net income retained equals 50% of the quarter's net income after taxes (assume the company pays 50% of its net income in dividends).
- CASH NEEDED TO START QUARTER: Assume that this equals 25% of the calculated assets invested in the division at the start of the quarter. Thus, as assets increase, cash required to run the business increases.
- UNITS SOLD—MOVING AVERAGE: Compute the forecasted Units Sold for these quarters, using a moving average. Enter a *5* for the interval size.
- UNITS SOLD—SIMPLE AVERAGE: Use the AVERAGE() function to compute the average of the first eight quarters. Thus, each of Quarters 9 to 12 will have the same forecasted units sold.
- UNITS SOLD—REGRESSION: Compute forecasted units sold using linear regression. The TREND() function would work well for this purpose.
- SELLING PRICE: The forecasted selling price depends on the economy. If a good economy is predicted, the selling price in a quarter will be 105% of the prior quarter's selling price. If the economy is OK, the selling price will remain unchanged from the prior quarter's selling price. If the economy is bad, the selling price will be 5% less than the prior quarter's selling price.
- UNIT VARIABLE COST: Due to expected inflation in raw materials and wages, each quarter's unit variable cost will be a dollar more than the prior quarter's cost.
- INTEREST RATE: If a Good economy is predicted, the quarter's interest rate on internally generated debt will be 10%. If the economy is OK, the interest rate will be 8%. If the economy is Bad, the rate will be 6%.
- RETURN ON ASSETS INVESTED: This ratio equals the net income after taxes in the quarter, divided by the assets invested at the start of the quarter.

INCOME STATEMENT AND NET CASH FLOW STATEMENT Section

The forecast for Net Income and Cash Flow starts with cash on hand at the beginning of the first period. This is followed by an income statement and concludes with a calculation of cash on hand at the end of the fourth quarter. Results in this section should be formatted for zero decimals (i.e., no pennies). Your spreadsheet sections should look like those shown in Figures 11-23 and 11-24. Line items are discussed after each figure.

	A	B	C	D	E	F
33	**INCOME STATEMENT AND CASH FLOW STATEMENT**		**9**	**10**	**11**	**12**
34	BEGINNING CASH ON HAND	**NA**				
35						
36	REVENUE (SALES)	**NA**				
37	VARIABLE COSTS	**NA**				
38	FIXED COSTS	**NA**				
39	TOTAL COSTS	**NA**				
40	INCOME BEFORE INTEREST AND TAXES	**NA**				
41	INTEREST EXPENSE	**NA**				
42	NET INCOME BEFORE TAXES	**NA**				
43	INCOME TAX EXPENSE	**NA**				
44	NET INCOME AFTER TAXES	**NA**				

Figure 11-23 INCOME STATEMENT AND CASH FLOW STATEMENT section

- BEGINNING CASH ON HAND: This is the cash on hand at the end of the *prior* quarter.
- REVENUE (SALES): Revenue is based on units sold and selling price per unit. The applicable units sold value depends on the units sold forecast input value (1, 2, or 3).
- VARIABLE COSTS: Variable costs are based on the number of units sold and the variable cost per unit.
- FIXED COSTS: In each quarter, this value is expected to be a million dollars more than the value for Quarter 8. The value will be the same in quarters 9 to 12.
- TOTAL COSTS: This is the total of variable costs and fixed costs.
- INCOME BEFORE INTEREST AND TAXES: This is revenue less total costs.
- INTEREST EXPENSE: This is based on the quarter's interest rate and the debt owed at the *beginning* of the quarter.
- NET INCOME BEFORE TAXES: This is income before interest and taxes, less interest expense.
- INCOME TAX EXPENSE: This expense is zero if income before taxes is zero or negative; otherwise, apply the tax rate to net income before taxes to calculate the income tax expense.
- NET INCOME AFTER TAXES: This is income before taxes less income tax expense.

	A	B	C	D	E	F
46	NET CASH POSITION (NCP) BEFORE BORROWING AND REPAYMENT OF DEBT (BEG YR CASH + NET INCOME)	**NA**				
47	ADD: INTERNAL BORROWING	**NA**				
48	LESS: INTERNAL REPAYMENT	**NA**				
49	EQUALS: ENDING CASH ON HAND	2000000				

Figure 11-24 END-OF-THE-QUARTER CASH ON HAND section

- Prior quarter values are mostly *NA*, except that $2 million cash was on hand to end the prior year, and is thus on hand to start Quarter 9.
- NET CASH POSITION (NCP) is the beginning of quarter cash plus net income after taxes in the period.

- ADD: INTERNAL BORROWING: If the NCP is less than the minimum, the division must borrow enough cash to get to the minimum cash required. If the NCP is greater than or equal to the minimum, then no cash is borrowed.
- LESS: INTERNAL REPAYMENT: If the NCP is more than the minimum required, the division can afford to pay back prior amounts owed to start the quarter, but not to take the division below the minimum cash required, of course.
- EQUALS: ENDING CASH ON HAND: End-of-quarter cash equals the NCP plus any borrowing and less any repayment in the quarter.

DEBT OWED Section

Your spreadsheet body ends with a calculation of debt owed at the end of the quarter, as shown in Figure 11-25. An explanation of line items follows the figure.

	A	B	C	D	E	F
51	DEBT OWED	NA	9	10	11	12
52	BEGINNING-OF-THE-QTR DEBT OWED	NA				
53	ADD: INTERNAL BORROWING	NA				
54	LESS: INTERNAL REPAYMENT	NA				
55	EQUALS: END-OF-THE-QTR DEBT OWED	2000000				

Figure 11-25 DEBT OWED section

- BEGINNING-OF-THE-QUARTER DEBT OWED: Prior quarter values are mostly *NA*, except that $2 million cash was owed by the division to end the prior year, and is thus owed to start Quarter 9.
- ADD: INTERNAL BORROWING/ LESS: INTERNAL REPAYMENT: Amounts borrowed and repaid have been calculated and can be echoed to this section.
- EQUALS: END-OF-THE-QTR DEBT OWED: The amount owed at the end of a quarter equals the amount owed to start the quarter, plus the amount borrowed, less any amount repaid.

Case 11

ASSIGNMENT 2 USING THE SPREADSHEET FOR DECISION SUPPORT

You will now complete the case by (1) using the spreadsheet to gather the data needed to decide whether the division's paintball division looks like a good business to be in, (2) documenting your recommendation in a memorandum, and (3) if your instructor specifies, an oral presentation.

Assignment 2A: Using the Spreadsheet to Gather Data

You have built the spreadsheet to forecast the division's operating results in the next four quarters. Units sold have been forecast in three different ways. Operating data have been forecast using three different economic scenarios.

World Wide Games management requires a 7% return on assets invested for its businesses. Those that cannot meet this minimum must be looked at critically, i.e., can the business be improved or should it be sold off?

Thus, WWG management (and the Master Blaster Division management) wants to know how the business forecast looks, versus the 7% benchmark, in the coming year. You will need to use the spreadsheet to see what scenarios (if any) result in acceptable returns and what scenarios (if any) result in unacceptable returns.

You run "what-if" scenarios with the nine sets of input values. You should manually enter the input value combinations. Note the Quarter 12 return on investment for each combination. For Quarter 12 values that are close to the 7% benchmark, note the return on investment value's trend, for Quarters 9 to 12.

Your instructor may require you to prepare charts for the data—a scatter plot and/or a linear-regression trend line.

When you are done gathering data, print the entire workbook (with any input value combination entered). Also, print any charts that your instructor requires. Then, save the spreadsheet (File—Save).

Assignment 2B: Documenting Your Recommendation in a Memorandum

Now you will document your recommendation in a memorandum. Open MS Word, and write a brief memorandum to management about forecasted operations. Here is guidance on your memorandum:

- Your memorandum should have a proper heading (DATE / TO / FROM / SUBJECT). You may wish to use a Word memo template (**File**—click **New**, click **On my computer** in the Templates section, click the **Memos** tab, choose **Contemporary Memo**, then click **OK**).
- In the first paragraph, tell which scenarios result in a return on assets greater than 7% and which do not. Support the statement graphically, including a table like the one shown in Figure 11-26.
- If a scenario has a return close to 7%, you should comment on the trend in the four quarters. Clearly, an upward trend towards a marginal Quarter 12 value would be preferable to a marginal Quarter 12 value that is at the end of a downward trend.
- If returns on investment differ significantly under the different Units Sold forecasts, you should try to put the forecasted Units Sold values in context. A scatter plot of the first eight quarter values may help put things in perspective. A highly dispersed "cloud" of data points would yield less-than-reliable regression values. Similarly, a scatter-plot cloud that is shaped like a boomerang would imply a non-linear relationship between time and sales—and, thus, the regression forecast might be unreliable. The R-squared value should help you assess the worth of the forecasted values. If forecasted values seem unreliable, you should suggest other kinds of research needed to improve the forecast.
- You are not required to forecast what kind of economy will prevail. Your findings should be in the form: "If the economy is good and the simple average forecast of units sold is accurate, the 12th quarter return on investment is: X%."
- Nevertheless, if it appears that the business can earn an acceptable return on investment only if there is a favorable economy and an optimistic sales forecast, then you should point that out to management. On the other hand, if it looks as if the business will succeed no matter what, you should point that out. If prospects seem unclear to you, you should state why you feel that way and support your opinion with the data.

Enter the table into Word, using the following procedure:

1. Select the **Table** menu option, point to **Insert**, then click **Table**.
2. Enter the number of rows and columns.

3. Select **AutoFormat** and choose **Table Grid 1.**
4. Select **OK**, and then select **OK** again.

Your table should resemble the format of the table shown in Figure 11-26.

Return on Assets (ROA) Forecast			
	Simple Average	**Moving Average**	**Regression**
Economy			
Good			
OK			
Bad			

Figure 11-26 Form of Return on Assets table to insert in memorandum

Assignment 3 Giving an Oral Presentation

Your instructor may request that you also present your results in an oral presentation. If so, assume that management wants wider knowledge of your findings. You have been asked to give a presentation explaining your results to the company's senior management. Prepare to explain your forecast and results to the group in 10 minutes or fewer. Use visual aids or handouts that you think are appropriate. Tutorial E has guidance on how to prepare and give an oral presentation.

Deliverables

1. Printouts of your memorandum
2. Spreadsheet printouts, including charts that your instructor has required
3. Disk, which should have your Word memo file and your Excel spreadsheet file

Staple the printouts together, with the memorandum on top. If there is more than one .xls file on your disk, write your instructor a note, stating the name of your model's .xls file.

12 CASE

The Social Safety Fund 75-Year Budget

DECISION SUPPORT USING EXCEL

PREVIEW

The government of a prosperous country operates a retirement program, called the Social Safety Fund, for the benefit of retiring citizens. Via payroll tax, citizens pay into the Fund throughout their employment. When citizens retire, they are paid a yearly income out of the Fund. However, a significant percentage of the country's work force is expected to retire in the next decade, and when that happens, it appears that Fund payments will exceed Fund receipts. Fund administrators and the country's political leaders need a long-term projection of the Fund's solvency. In this case, you will use Excel to make a 75-year operating budget for the Fund, in which you will forecast yearly receipts, expenditures, and Fund balances.

PREPARATION

- Review spreadsheet concepts discussed in class and/or in your textbook. This case requires that you know how to use the Excel functions IF(), SUM(), and VLOOKUP(). The case also requires that you know how to transfer data between worksheets, use basic charting techniques, name cells, and refer to named cells in formulas.
- Complete any exercises that your instructor assigns.
- Complete any part of Tutorial C that your instructor assigns, or refer to it as necessary.
- Review file-saving procedures for Windows programs. These are discussed in Tutorial C.
- Refer to Tutorial E as necessary.

BACKGROUND

The country began the Social Safety Fund plan in 1935, after the Great Depression. For almost a decade, the Depression ravaged the country's economy. The unemployment rate exceeded 20% year after year, and the gross domestic product shrank. Large segments of the population experienced real poverty, including many older people of retirement age.

To reduce economic hardship on older Americans, the citizenry, acting through its democratically elected political leaders, adopted a social safety net called the "Social Safety Act." Under this act, working people agreed to have their wages and salaries taxed in order to establish a fund. Out of this fund, retirees would be paid an amount sufficient to stave off poverty. Thus, the country's younger people agreed to take care of its older people. It was understood that the fund would survive in perpetuity, so the sacrifices made by citizens when they were young would be recouped when they retired.

Since then, the country has grown greatly in population and national wealth, and the Social Safety Fund has grown with it. Annual tax receipts paid into the Fund have almost always exceeded amounts paid out to retirees—sometimes by a wide margin. For decades, the country's senior citizens have been secure in the knowledge that Fund payments, plus private pensions and savings, would be sufficient to support a comfortable retirement.

As the 21st century gets underway, however, these demographic, economic, and political factors threaten the Fund's long-term financial stability:

- Although the economy continues to grow overall, this increase is not always matched by wage increases for people paying into the Fund. For a number of years, the average person's pay has remained at about $30,000 per year.
- The payroll tax supporting the Fund has remained at 12.4% for many years (6.2% paid by the worker and 6.2% by the employer). The current political climate is anti-tax. Raising the tax rate to increase Fund revenues would be unpopular with workers and employers—and elected officials who count on their votes.
- Families are not as large as they once were. The number of births in the country no longer far exceeds the number of deaths. In the very long term, the number of those entering the workforce each year will more or less equal the number leaving.
- In the three decades following World War II, which ended in 1945, birth rates were quite high. People born in this era, known as "the war babies," paid huge sums of money into the Fund for many years. This large portion of the population is now about ready to retire and draw money from the Fund. In addition, they will be drawing from it for a long time because people now live much longer than the Fund's founders anticipated. The average citizen now lives well into their 70s.

Thus, because of these factors, in the future, Fund payments will probably start to exceed Fund receipts, perhaps by a wide margin.

Fund administrators and the country's elected officials know that the Fund must be kept on a sound financial footing. If retirees and soon-to-be retirees see their retirement threatened, they may vote for new political leadership. Similarly, young people do not want to pay into a Fund that they think will not be solvent when they retire. Accordingly, there are economic and political needs for an accurate long-term financial projection of the Fund's financial health. If changes are needed, it will be better to make them sooner, when changes are affordable, rather than later, when they are not so affordable.

You have been called in to do the projection in Excel. Your initial reaction might be that the project is not very urgent or very important. You might ask, "What is there to worry about? The country has built up huge cash reserves for decades. Now the country will have to pay out of reserves for a while. Isn't that what savings are for?"

To appreciate the answer to this question, more elaboration on Fund finances is needed. *Example*: Assume (using illustrative small numbers) that the Fund has a $5,000 balance to start a year. Suppose that during one year, $1,000 is paid into the Fund and $500 is paid out in retirement benefits. At the end of the year, the Fund should have $5,500 in cash in the Treasury ($5,000 + ($1,000 - $500)). Right?

Well, not exactly. There are two reasons why it won't have $5,500 in cash at the Treasury:

1. The Fund's balance is not actually held as cash in a bank. The taxes collected are given to the Treasury. In return, the Treasury issues special non-negotiable bonds to the Fund—in effect, the Treasury gives the Fund an IOU each year in which receipts exceed payments (which is most years). The Treasury then uses the surplus to pay the country's other obligations, such as those for defense, welfare, schools, national parks, and so on.
2. The special Treasury Bonds carry a 7% interest rate. Thus, $5,500 is not the correct balance. The Fund would earn 7% on the balance at the beginning of the year—in the example, $350 ($5000 * .07). The correct Fund balance is $5000 + $350 + $1,000 – $500 = $5,850.

Perhaps you can now see why there is such concern over the Fund's stability. There is a huge Fund reserve, *but the amount is not held in cash.* It is held in IOUs, which cannot be sold by the Fund because they are non-negotiable. The Fund may have a large balance, but it cannot be accessed to be spent.

If Fund managers must draw on its reserves to pay benefits, they will have to go to the Treasury and ask for real cash, which could be provided in one of three ways:

- The Treasury could go to the lawmakers and ask for a general tax increase.
- The Treasury could ask for authority to issue more general-purpose Treasury Bonds in the open market (as it so often does to finance government business).
- The Treasury could just print the money, which would be, by definition, inflationary.

None of these things should happen on an *emergency* basis in a well-run democracy. Thus, some financial planning is in order; hence, the need for your projection in Excel.

Assignment 1 Creating a Spreadsheet for Decision Support

In this assignment, you will produce a spreadsheet that models Fund operations for 75 years, 2005–2079. Then, in Assignment 2, you will write a memorandum to Fund management that explains your findings. In Assignment 3, you will prepare an oral presentation of your analysis.

Next, you will create the projection in Excel. You will be given some hints on how each worksheet and section should be set up before entering cell formulas. Your spreadsheet should contain four worksheets:

- **SUMMARY OF REVENUES, EXPENDITURES, AND FUND BALANCES**
- **FORECASTED REVENUES**
- **FORECASTED EXPENDITURES**
- **FORECASTED POPULATION FACTORS**

A discussion of each worksheet follows. *The spreadsheet skeleton is available to you, which you should get from your instructor.* To access the spreadsheet skeleton, go to your Data Files. Select Case 12, then select **SOCIAL.xls**.

SUMMARY OF REVENUES, EXPENDITURES, AND FUND BALANCES Worksheet

Your SUMMARY OF REVENUES, EXPENDITURES, AND FUND BALANCES worksheet should have the structure shown in Figure 12-1:

	A	B	C	D	E	F	G
1	SUMMARY OF REVENUES, EXPENDITURES, AND FUND BALANCES						
2							
3	CONSTANTS						
4	TAX RATE (.XXX)	0.124					
5	INTEREST RATE (.XX)	0.07					
6	CHANGE IN BENEFITS RATE (+/- .XX)	0.01					
7	WORKER INCOME GROWTH (.XX)	0					
8	NUMBER OF YEARS BENEFITS DELAYED (X)	0					
9							
10		2004	2005	2006	2007	2008	2009
11	FUND INCOME	NA					
12							
13	FUND EXPENDITURES	NA					
14							
15	DIFFERENCE: SURPLUS <DEFICIT>	NA					
16							
17	FUND BALANCE -- BEGINNING OF YEAR	NA					
18							
19	FUND BALANCE -- END OF YEAR	1683.7					
20							
21							
22			CHARTING VALUES				
23	*(EST INCOME IN BILLIONS. E.G.,*				2010	2020	2030
24	*492.9 = 492,900,000,000. FUND BALANCE*		FUND BALANCE				
25	*IN BILLIONS. E.G., 1683.7 =*						
26	*$1,683,700,000,000*		RECEIPTS LESS				
27			PAYMENTS				
28							
29			RATIO OF				
30			CONTRIBUTORS TO				
31			RECIPIENTS				

Figure 12-1 Structure of the SUMMARY OF REVENUES, EXPENDITURES, AND FUND BALANCES worksheet

The sections of the worksheet are discussed next.

CONSTANTS Section

- TAX RATE: The tax rate on payrolls is 12.4%. Thus, .124 * $30,000 is now contributed each year for the average worker.
- INTEREST RATE: The interest rate on Treasury Bonds is 7% a year.
- CHANGE IN BENEFIT RATE: The average benefit paid to a retiree has been increasing 1% each year for some time. This rate of increase is expected to continue.
- WORKER INCOME GROWTH: This is the growth rate for worker pay. If it is assumed to be zero, the average pay for a worker will be $30,000.
- NUMBER OF YEARS BENEFITS DELAYED: Currently, a worker can retire at 62 and start collecting benefits. Lawmakers could delay this to reduce expenditures.

You should name each CONSTANT cell. CONSTANT section values will be echoed to other worksheets and used there. When you do that you refer to the cell by its name, not by its cell address.

Note that the user could play "what-if" with the constants, in effect turning them into inputs:

- The tax rate could be increased to increase fund revenues.
- The interest rate could be changed: Lawmakers might specify a lower rate to save some money. The rate can be set to zero to see what revenues are, from payroll taxes alone.
- Lawmakers might specify higher or lower benefits in future years.
- The average worker income could be increased one year to the next, perhaps to reflect assumed productivity increases in the workplace that would actually raise workers' pay.
- Lawmakers could say that people must wait until age 63 (or 64, or whatever). The change would be reflected in the NUMBER OF YEARS BENEFITS DELAYED value.

Spreadsheet Body

- FUND INCOME and FUND EXPENDITURES for each year are echoed here from other worksheets.
- DIFFERENCE is income less expenditures. If positive, there is a surplus in the year that increases the Fund balance. If negative, a deficit that decreases the Fund balance.
- FUND BALANCE—BEGINNING OF YEAR: This is the balance at the beginning of a year. Of course, this equals the balance at the end of the prior year.
- FUND BALANCE—END OF YEAR: This equals the beginning of year balance, plus the difference in the year. The balance at the end of 2004 is shown.

There will be columns for 75 years (2005–2079). Only a few years are shown in the figure, due to space restrictions.

Charting Values

You will need to chart some of your results. In this area, you should echo key results for charting. Years 2010, 2020, 2030, 2040, 2050, 2060, 2070, and 2079 are assumed. (Your instructor may ask you to add other years.)

- FUND BALANCE: Chart the Fund balance at 2010, 2020, ... , 2070, 2079.
- RECEIPTS LESS EXPENDITURES: Show the difference between tax receipts and benefits payments (i.e., use actual cash inflows and cash outflows on other sheets; do not include interest value). In a year in which this number turns negative, the Fund will need to ask the Treasury for help!
- RATIO OF CONTRIBUTORS TO RECIPIENTS: How many people are paying into the Fund versus those receiving benefits from it? This ratio is expected to decline due to the "war baby" effect. This ratio is the quotient of values computed in two other worksheets.

FORECASTED REVENUES Worksheet

The structure of the FORECASTED REVENUES worksheet is shown in Figure 12-2. Line items are discussed after the figure.

	A	B	C	D	E	F	G
1	**FORECASTED REVENUES**						
2							
3	**CONSTANTS**						
4	TAX RATE (.XXX)	0.124					
5	INTEREST RATE (.XX)	0.07					
6	CHANGE IN BENEFITS RATE (+/- .XX)	0.01					
7	WORKER INCOME GROWTH (.XX)	0					
8	NUMBER OF YEARS BENEFITS DELAYED (X)	0					
9							
10		**2004**	**2005**	**2006**	**2007**	**2008**	**2009**
11	NUMBER OF CONTRIBUTING WORKERS	**NA**					
12							
13	AVERAGE TAXABLE INCOME	30.0					
14							
15	TOTAL RECEIPTS	**NA**					
16							
17	INTEREST INCOME	**NA**					
18							
19	TOTAL INCOME	**NA**					

Figure 12-2 Structure of the FORECASTED REVENUES worksheet

CONSTANTS Section

The values in the CONSTANTS section (cells 4–8) are echoed from the SUMMARY OF REVENUES, EXPENDITURES, AND FUND BALANCES worksheet. Use "exclamation point" notation. *Example*: If the SUMMARY OF REVENUES, EXPENDITURES, AND FUND BALANCES worksheet is named *Sheet1*, then echoing a value from a cell named *MyCell* in *Sheet1* would be shown thus: *=Sheet1!MyCell*.

Spreadsheet Body

- NUMBER OF CONTRIBUTING WORKERS: The value for each year is echoed from the FORECASTED POPULATION FACTORS worksheet.
- AVERAGE TAXABLE INCOME: This is $30,000 to start. The amount is increased by the value of worker income growth, year to year. *Example*: If the growth rate is 1%, then the value for 2005 would be $30,300. The value in 2006 would be $30,603, and so on for all years.
- TOTAL RECEIPTS: This is based on the number of contributors, the average taxable income in a year, and the tax rate.
- INTEREST INCOME: This is based on the interest rate (a constant) and the Fund balance at the start of the year. That balance is shown on the SUMMARY OF REVENUES, EXPENDITURES, AND FUND BALANCES worksheet. Notice that there is interest income only if the Fund balance is positive.
- TOTAL INCOME: This is the sum of receipts and interest.

FORECASTED EXPENDITURES Worksheet

The structure of the FORECASTED EXPENDITURES worksheet is shown in Figure 12-3. Line items are discussed after the figure.

	A	B	C	D	E	F	G
1	FORECASTED EXPENDITURES						
2							
3	CONSTANTS						
4	TAX RATE (.XXX)	0.124					
5	INTEREST RATE (.XX)	0.07					
6	CHANGE IN BENEFITS RATE (+/- .XX)	0.01					
7	WORKER INCOME GROWTH (.XX)	0					
8	NUMBER OF YEARS BENEFITS DELAYED (X)	0					
9							
10		2004	2005	2006	2007	2008	2009
11	NUMBER OF BENEFIT RECIPIENTS	NA					
12							
13	AVERAGE BENEFIT PAYMENT	9,800					
14							
15	TOTAL BENEFIT PAYMENTS	NA					
16							
17	INTEREST PAYMENTS	NA					
18							
19	TOTAL EXPENDITURES	NA					
20							

Figure 12-3 Structure of the FORECASTED EXPENDITURES worksheet

CONSTANTS Section

The CONSTANTS section is echoed from the SUMMARY OF REVENUES, EXPENDITURES, AND FUND BALANCES worksheet. Use "exclamation point" notation.

Spreadsheet Body

- NUMBER OF BENEFITS RECIPIENTS: The value for each year is echoed from the population worksheet.
- AVERAGE BENEFIT PAYMENT: This amount was $9,800 in 2004. The amount is expected to increase by the CHANGE IN BENEFITS RATE, a constant.
- TOTAL BENEFIT PAYMENTS: This is based on the number of recipients and the average payment.
- INTEREST PAYMENTS: If the Fund balance goes negative, the Fund is assumed to be in a borrowing position, and it will have to pay interest on borrowings. This amount is based on the interest rate (a constant) and the Fund balance at the start of the year, if the balance is negative. The Fund balance is shown on the SUMMARY OF REVENUES, EXPENDITURES, AND FUND BALANCES worksheet.
- TOTAL EXPENDITURES: This is the sum of benefits and interest payments.

FORECASTED POPULATION FACTORS Worksheet

The structure of the FORECASTED POPULATION FACTORS worksheet is shown in Figure 12-4. Line items are discussed after the figure.

	A	B	C	D	E	F	G
1	**FORECASTED POPULATION FACTORS**						
2							
3	**CONSTANTS**						
4	TAX RATE (.XXX)	0.124					
5	INTEREST RATE (.XX)	0.07					
6	CHANGE IN BENEFITS RATE (+/- .XX)	0.01					
7	WORKER INCOME GROWTH (.XX)	0					
8	NUMBER OF YEARS BENEFITS DELAYED (X)	0					
9							
10		**2004**	**2005**	**2006**	**2007**	**2008**	**2009**
11	NUMBER OF CONTRIBUTING WORKERS	**NA**					
12							
13	NUMBER OF BENEFIT RECIPIENTS	**NA**					
14							
15	**EXPECTED POPULATION VALUES**						
16				**RECIP-IENTS**		**CONTRIB-UTORS**	
17	*(EST POP IN MILLIONS. E.G., 290.80 =*	**YEAR**	**EST POP**	**% >= 62**	**% 0 - 20**	**% PAYING**	
18	*290,800,000)*	2004	290.81	0.1467	0.2934	0.5599	
19		2005	294.66	0.1532	0.2892	0.5576	
20		2006	297.91	0.1582	0.2858	0.5560	

Figure 12-4 Structure of the FORECASTED POPULATION FACTORS worksheet

CONSTANTS Section

This section is echoed from the SUMMARY OF REVENUES, EXPENDITURES, AND FUND BALANCES worksheet. Use "exclamation point" notation.

EXPECTED POPULATION VALUES Section

These values are at the bottom of the spreadsheet in a "lookup table." (The values are discussed at this point to make the meaning of the rest of the worksheet's values more clear.) You can assume that the government's Census Bureau has provided data as follows:

- EST POP...: This is the expected population of the country in the years 2004 to 2079.
- RECIPIENTS % ≥ 62: This is the percentage of citizens who are age 62 or greater—the people who are assumed to be recipients.
- % 0 – 20: This is the percentage of citizens who are 0–20 years old. These people are assumed to be not yet contributors to the Fund.
- CONTRIBUTORS % PAYING: This is the percentage of people who are paying into the Fund—all those who are not recipients and over 20 years old.

Spreadsheet Body

Notice that these data are shown for years 2004 through 2009—only the first few years are shown here, to save space. The body of the spreadsheet has these values:

- NUMBER OF CONTRIBUTING WORKERS. This amount is based on the percentage of the population that is contributing and the total population. The percentage is increased by 1.5% for each year that benefits are delayed—the delay value is a constant. If the constant is a *1*, then *63* is the retirement age, and the value of the percentage of contributors in the table must be incremented by 1.5% in your formula. *Note*: You should access the contributor's percentage paying and estimated population values in the *lookup table* by using the VLOOKUP() function.
- NUMBER OF BENEFIT RECIPIENTS: This amount is based on the percentage of the population that is receiving benefits and the total population. The percentage is decreased by 1.5% for each year that benefits are delayed. If the constant is a *1*, then

Case 12

63 is the retirement age, and the value of the percentage of recipients in the table must be decremented by 1.5% in your formula. Note: You should access the recipient percentage and the estimated population values in the *lookup table* by the VLOOKUP() function.

Notice that the values for the contributing workers and the benefit recipients are echoed to other worksheets.

ASSIGNMENT 2 USING THE SPREADSHEET FOR DECISION SUPPORT

You will now complete the case by (1) using the spreadsheet to gather the data needed to decide if and when the Fund faces serious deficits, and what actions might be taken to improve the Fund's financial status; (2) documenting your recommendation in a memorandum; and (3), if your instructor specifies, an oral presentation, given either individually or with a team of classmates.

Assignment 2A: Using the Spreadsheet to Gather Data

You have built the spreadsheet to forecast the Fund's operating results in the next 75 years.

You should use the results to gather data to answer certain questions. Assume default constants are a .124 tax rate, .07 interest rate, .01 increase in benefits, 0 income growth, and 0 years benefits delayed.

- What is the 2079 Fund balance with default constants?
- Does the Fund balance turn negative? If so, in what year?
- Do expenditures ever exceed receipts? If so, in what year?

You should use the model to play what-if with the results. Here are the rules for changing constants that are—and are NOT—allowable:

- Tax rate cannot increase.
- Interest rate cannot increase.
- Benefits need not be increased, but they cannot be decreased (i.e., the increase value can be zero).
- Income growth cannot exceed .01 (assume a 1% real income growth for 75 years would be considered quite good by most economists).
- Benefits can be delayed by one year, but no longer than one year.

Fund management wants to know how to get the 2079 balance to be a positive number. What reasonable combinations of constants would allow that, if any?

Fund management would be very happy if a receipts-expenditures deficit could be eliminated or made "small." What reasonable combinations of constants would allow that throughout all years, if any? (Here, $150–200 billion is a small number to have to borrow in a year.)

To answer these questions, run "what-if" scenarios with the constant values. Manually enter the values. Gather data that you need on a separate sheet of paper. Some of this data should be recorded in the Charting data area. (Your instructor may change Charting requirements.) You should create these charts:

- Fund balances over time (time values are on the X axis, and Fund balances are on the Y axis).
- Surplus or deficit over time (time values are on the X axis, and the differences between cash in and cash out are on the Y axis).

- Contributors-to-recipients ratio over time (time values are on the X axis, and ratios are on the Y axis).

When you are done gathering data, print the entire workbook (with any input value combination entered). Then, save the spreadsheet.

Assignment 2B: Documenting Your Recommendation in a Memorandum

Open MS Word, and write a memorandum to Fund management about forecasted operations. Your memorandum should have the following elements:

- A proper heading (DATE / TO / FROM / SUBJECT). You may wish to use a Word memo template (**File**—click **New**, click **On my computer** in the Templates section, click the **Memos** tab, choose **Contemporary Memo**, then click **OK**).
- Devote a section to each of management's questions and concerns.
- Support your findings graphically, using the charts required and/or others that you think are pertinent.
- Take a stand. How much trouble is the Fund in? Are there practical solutions—politically and economically viable ways to make the Fund solvent in the long term? Keep in mind that the numbers here are very large—an amount such as a $100-billion shortfall is not an amount that would doom the system.

If you need to support your findings with tables in your MS Word document, enter the table into Word, using the following procedure:

1. Select the **Table** menu option, point to **Insert**, then click **Table**.
2. Enter the number of rows and columns.
3. Select **AutoFormat** and choose **Table Grid 1**
4. Select **OK**, and then select **OK** again.

Assignment 3 Giving an Oral Presentation

Your instructor may request that you also present your results in an oral presentation. If so, assume that Fund management wants wider knowledge of your findings. They have asked you to give a presentation to the president of the country and her cabinet, explaining your results. Prepare to explain your projection methods and results to the group in 10 minutes or fewer. Use visual aids or handouts that you think are appropriate. Tutorial E has guidance on how to prepare and give an oral presentation.

Deliverables

1. Printout of your memorandum
2. Spreadsheet printouts, including charts
3. Disk or CD, which should have your Word memo file and your Excel spreadsheet file

Staple the printouts together, with the memorandum on top. If there is more than one .xls file on your disk or CD, write your instructor a note, stating the name of your model's .xls file.

13
CASE

Challenge Case: The Warehouse Location Decision

DECISION SUPPORT USING EXCEL

PREVIEW

Fun Phones make phones at two locations and leases warehouses at both locations. The warehouse leases expire at the end of 2005. Management has decided to build their own warehouses rather than continue leasing them. In this case, you will use Excel to make a net income and cash flow forecast, which will help Fun Phone's management decide where to build the new warehouses.

Unlike previous cases, you will not receive guided instruction in how to set up spreadsheet sections. Instead, you will need to review the information provided and determine the form and content of your spreadsheet and final output.

PREPARATION

- Review spreadsheet concepts discussed in Tutorials C and D and in prior cases.
- Refer to Tutorial E as necessary.
- Review file-saving procedures for Windows programs discussed in Tutorial C.

BACKGROUND

In recent years, Americans have fallen in love with the cell phone. A number of companies now offer nifty phones that do email, photography, Internet commerce—and they even let you make phone calls!

Most of the cell phone companies do not actually make the phones that they sell you. Other companies make the phones for the phone companies. The phone companies do the final packaging and marketing to the general public.

Fun Phones is one of the companies that makes phones for cell phone companies. Fun Phones makes two very successful lines of phones, which they call the Ding-a-lingo and the Buzzz-me. (Of course, the cell phone companies re-name these in their marketing to the general public.) After only a few years on the market, these phones have quickly gained popularity with phone companies.

Fun Phones manufactures phones at its St. Louis, Missouri, and Boston, Massachusetts, production sites. The company has been leasing warehouse space in those cities as well. The phones are stored in the warehouses until they are shipped to cell phone companies.

The warehouse leases in both cities expire at the end of 2005, and the warehouse owners plan to increase warehouse rents exorbitantly. Rather than renew the leases, Fun Phones management has decided to build its own warehouses in other cities.

Fun Phones management has identified three possible warehouse locations: Portland, Oregon; Phoenix, Arizona; and Atlanta, Georgia. Fun Phones would, however, construct warehouses at only *two* of these sites. You have been hired as a consultant to analyze the proposed locations and to help your client determine which of the two would be financially the best. You set a meeting with your client to gain an overall picture of their operation. You plan to find out about the following things in your meeting: (1) factors that are known; (2) factors that are not known; and (3) factors that must be calculated by spreadsheet model.

Known Factors

No matter where the warehouses are built, certain factors are known. These factors will either remain the same from year to year, or their degree of change can be predicted with accuracy. These factors are discussed next.

In the next three years (2006, 2007, and 2008), Fun Phones management expects to sell all the phones that they produce. Production of the two kinds of phones will be divided evenly between the two manufacturing locations. Figure 13-1 shows the expected production (and sales) levels in the next three years.

Phone/Manufacturing Site	2006	2007	2008
Ding-a-lingo—St. Louis	500,000	750,000	1,000,000
Buzzz-me—St. Louis	750,000	850,000	1,000,000
Ding-a-lingo—Boston	500,000	750,000	1,000,000
Buzzz-me—Boston	750,000	850,000	1,000,000

Figure 13-1 Number of phones to be manufactured in each city

To reduce distribution and storage risks (from strikes, natural disasters, etc.), management plans to send half of each plant's production to each of the two selected warehouse sites. For example, if Atlanta and Phoenix are selected as warehouse sites, then half of Boston's Ding-a-lingo production will be sent to Atlanta and half will be sent to Phoenix; similarly, half of Boston's Buzzz-me production will be sent to Atlanta and half to Phoenix. St. Louis' phone production would be similarly split between the Atlanta and Phoenix warehouses.

Shipping costs will differ from each plant to each of the three proposed warehouse sites. Also, storage costs at each of the three warehouse sites will differ. Management has analyzed how much it will cost to ship and store a phone for each of the factory-to-warehouse combinations. Figure 13-2 shows the shipping and storage costs for each proposed city for the next three years.

From Factory Location	To Warehouse Location	Year 2006	Year 2007	Year 2008
St. Louis	Portland	4.50	4.95	5.45
St. Louis	Atlanta	5.50	6.05	6.66
St. Louis	Phoenix	3.50	3.85	4.24
Boston	Portland	6.00	6.60	7.26
Boston	Atlanta	4.50	4.95	5.45
Boston	Phoenix	5.50	6.05	6.66

Figure 13-2 Cost to ship and store one phone at each proposed warehouse location

In addition to the previously mentioned known factors, there are some additional known factors:

- The yearly tax rate will increase from year to year. For 2006, it will be 29%; for 2007, it will be 30%; for 2008, it will be 31%. The rate applies to net income before taxes.
- The company wants to have at least $10,000 in cash at the beginning of each year, 2006–2008. The company's banker will lend whatever is needed at the end of a year to begin the next year with $10,000.
- There will be fixed administrative costs each year. Those costs will be $1,200,000 each year, 2006–2008.

Unknown Factors

There are two key factors that are not known. In the next few years, the state of the economy is not known, but management sees no chance of an economic downturn. They think that the economy will either be unchanged or will inflate. The other unknown factor is the warehouse location that will *not* be chosen.

For each of the three years, 2006–2008, management wants to know (1) the net income after taxes, (2) the end-of-year cash on hand, and (3) the amount owed to the bank at the end of the year.

Amounts to Calculate

In 2005, it costs $5.00 to manufacture and market a phone made in St. Louis; it costs $5.50 for one made in Boston. Production and marketing costs are expected to rise in the next three years. Because the unemployment rate in St. Louis is lower than Boston's, management thinks wage rates and other costs would go up relatively more in St. Louis if the economy inflates. Figure 13-3 shows the expected year-to-year rates of increase in manufacturing and marketing costs, according to the production site and the state of the economy.

	Economy Is Unchanged	Economy Inflates
Phone made in St. Louis	1% per year	15% per year
Phone made in Boston	2% per year	5% per year

Figure 13-3 Expected year-to-year increase in manufacturing and marketing costs

The amount that Fun Phones can charge telephone companies for a cell phone depends on the state of the economy. In 2005, a Ding-a-lingo phone sells for $11.25, and the slightly more popular Buzzz-me phone sells for $12.25. If the economy remains unchanged, the selling prices of Ding-a-lingo and Buzzz-me phones are not expected to change in 2006–2008. But if the economy inflates, selling prices of each phone would rise 4% from one year to the next.

Interest must be paid on debt owed to the bank. If the economy remains unchanged, the interest rate will be 8%, based on debt owed at the beginning of the year. If the economy inflates, the rate will be 10%.

You must know how many of each kind of phone gets shipped from the manufacturing site to a storage site so you can compute shipping and storage costs. If a warehouse site is not selected then, of course, no phones are shipped there; otherwise, a factory's production is divided evenly between the two selected sites.

Examining the Client's Books

Prior to beginning the construction of your spreadsheet, you ask Fun Phones for a look in their books. You are interested in looking at the 2005 income statement and cash flow. There were no surprises, but you did note that the end-of-year cash on hand for 2005 will be $10,000, as previously noted, and that $3,000,000 of debt will be owed to the bank at the end of 2005.

The company's treasurer tells you that any cash on hand at the end of a year that exceeds the minimum cash required should be used to repay any debt owed. But such payments should not reduce the cash on hand below the minimum, of course.

As you conclude your meeting with your client, you agree that you will provide them with a spreadsheet model that indicates the best two-warehouse combination, given either an unchanging economy or an inflated economy. For each two-warehouse combination and state of the economy, your client wants to know their cash position at the end of 2008, debt owed at the *end* of 2008, and net income after taxes for 2008; those are the data that your spreadsheet should provide.

Assignment 1 Creating a Spreadsheet for Decision Support

In this assignment, you will produce a spreadsheet that models the business decision, as previously outlined. Your instructor will tell you if your work is to be done in one step or in two steps.

If your instructor tells you to do this assignment in one step, first, determine what spreadsheet sections you will need to set up. Also, determine the appropriate Excel tools to employ. Then make the necessary calculations. When you are finished, print out your results. Save your file one last time (File—Save). A good file name would be WAREHOUSE.xls. Then exit the file.

If your instructor tells you to do this assignment in two steps, you should first submit a spreadsheet design to your instructor. Set up a skeleton of the spreadsheet, including calculations and spreadsheet body line items. When you are finished, print out your spreadsheet. Save your file (File—Save). A good file name would be WAREHOUSE.xls. Then exit the file. When your instructor has approved the design, you enter the formulas for the calculations and spreadsheet body—in other words, you finish the spreadsheet. Save your file one last time (File—Save). Then exit the file.

ASSIGNMENT 2 DOCUMENTING YOUR RECOMMENDATIONS IN A MEMORANDUM

Write a memorandum to Fun Phones' president that explains your recommendation. Your memorandum should have a proper heading and format. The body of your memo should advise her which warehousing combination yields the best financial results given (1) an unchanging economy, and (2) an inflated economy. Support your recommendation with one or more tables that summarize your findings.

ASSIGNMENT 3 GIVING AN ORAL PRESENTATION

The president of Fun Phones is so impressed with your work, she has asked you to make a presentation to the board of directors about your analysis and findings. Each board member knows the business situation, so you do not need to provide much background information. Use appropriate graphics to support your recommendation.

DELIVERABLES

1. Printout of your memorandum, if required
2. Spreadsheet printouts
3. Disk or CD, which should have your Word memorandum file and your Excel spreadsheet file.

Staple the printouts together, with the memorandum on top. If there are other .xls files on your disk or CD, write your instructor a note, stating the name of this case's .xls file.

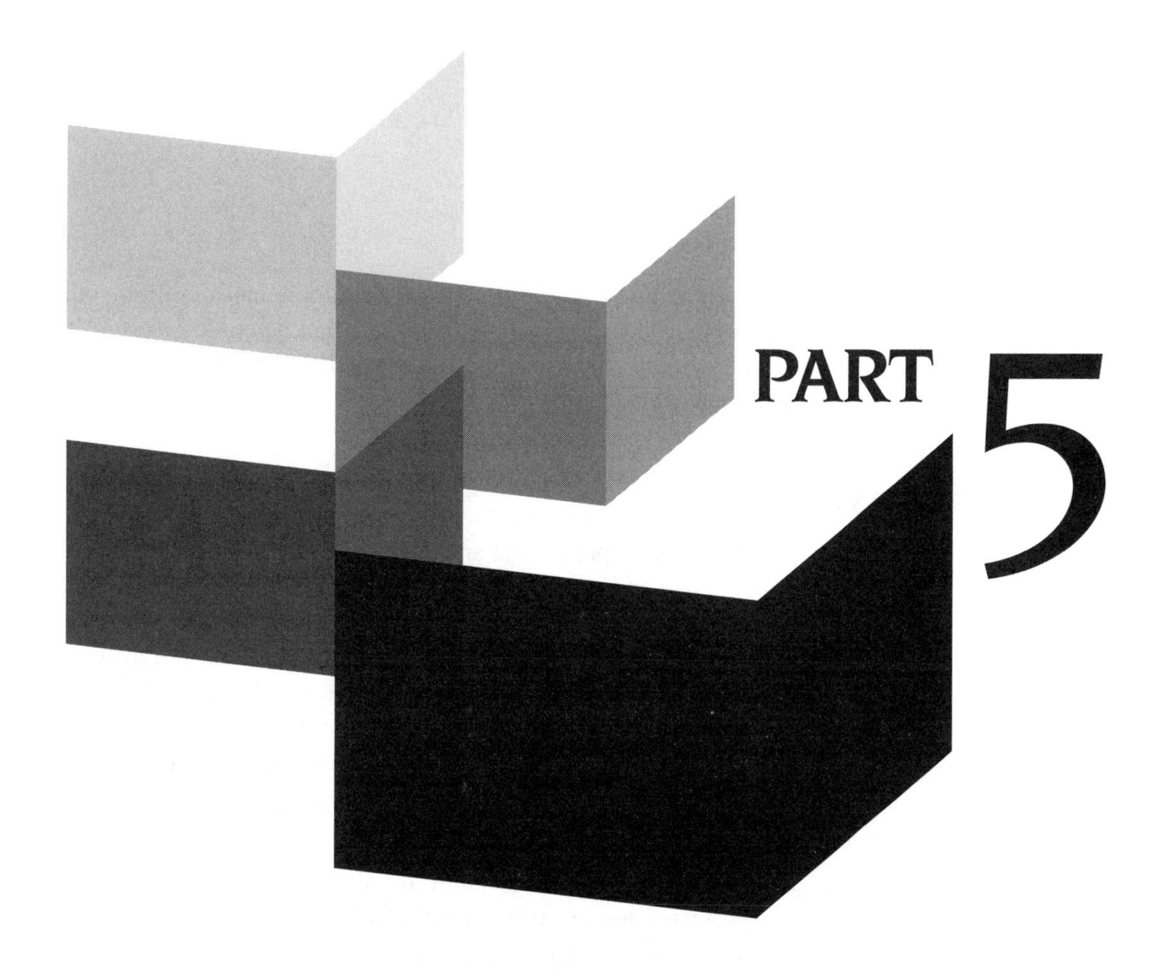

PART 5

Integration Case: Using Access and Excel

14 Your Uncle's Portfolio Construction Program

DECISION SUPPORT USING ACCESS AND EXCEL

PREVIEW

Your uncle just won the state lottery and now has a million dollars to invest. There are a number of stocks that he could invest in, and he has asked you to make a spreadsheet for him that would automate his selection process.

PREPARATION

- Review spreadsheet and database concepts discussed in class and/or in your textbook. This case requires you to use these Excel formulas: SUM(), IF(), COUNTIF(). The case requires you to understand these Excel features: transferring data between worksheets, charting, trend lines in charting, pivot tables, and conditional formatting.
- Complete any exercises that your instructor assigns.
- Obtain the database file STKLOOK.mdb from your instructor.
- Review any part of Tutorials A, B, C, or D that your instructor specifies, or refer to them as necessary.
- Review file-saving procedures for Windows programs. These are discussed in Tutorial C.
- Refer to Tutorial E as necessary.

BACKGROUND

Your uncle has always been a lucky fellow! A few months ago, he bought a lottery ticket and won. After state and federal taxes and rounds of serious family partying, your uncle still has a million dollars left. He wants to invest this money in the common stocks of well-known companies, subject to some simple investment rules. He wants a spreadsheet that will let him put together a model of his portfolio before he actually buys the stocks. Knowing that you know how to work with Access and Excel, he has asked you for help.

Your uncle has told you which stocks he would consider buying. He has asked you to gather recent market data and financial data about them and to put the data into a database. You have done this. The database tables are discussed next.

STOCKS Table

The STOCKS table shows permanent data about the candidate stocks. The data is shown in Figure 14-1. A discussion of the table's fields follows the figure.

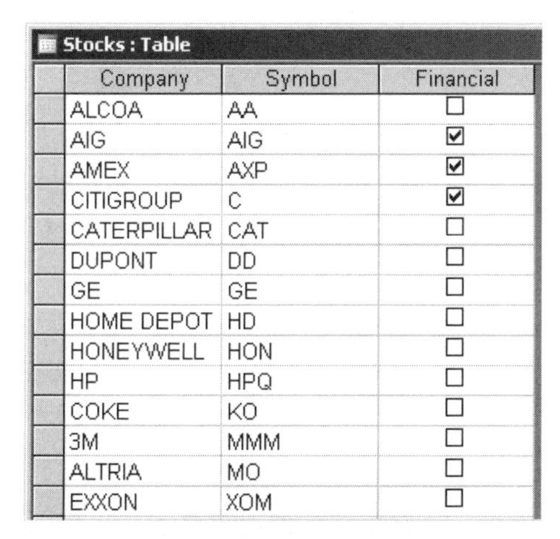

Stocks : Table

Company	Symbol	Financial
ALCOA	AA	☐
AIG	AIG	☑
AMEX	AXP	☑
CITIGROUP	C	☑
CATERPILLAR	CAT	☐
DUPONT	DD	☐
GE	GE	☐
HOME DEPOT	HD	☐
HONEYWELL	HON	☐
HP	HPQ	☐
COKE	KO	☐
3M	MMM	☐
ALTRIA	MO	☐
EXXON	XOM	☐

Figure 14-1 The STOCKS table

- Company—The Company field shows the company's name.
- Symbol—The Symbol field shows the company's stock exchange symbol.
- Financial—The Financial field shows whether the company's business is primarily financial. This is a *Yes/No* field. The check mark (for *Yes*) indicates that the company's business is primarily financial. (*Note*: *Yes/No* in Access is synonymous with *True/False* in logic, and sometimes a *Yes/No* field is referred to as a *True/False* field).

ACTIVITY Table

The ACTIVITY table's fields, shown in Figure 14-2, show market price activity for each stock. This activity is for the first trading day of each of seven past months. Partial data are shown. A discussion of the table's fields follows the figure.

Activity : Table

Symbol	Price Date	Open	High	Low	Close	Volume
AA	02-Jan-05	38	39.44	32.6	34.18	7465935
AA	02-Feb-05	34.22	38.15	33.36	37.47	5233368
AA	01-Mar-05	37.67	38.58	32.63	34.69	5160043
AA	01-Apr-05	34.67	36.6	30.5	30.75	7615633
AA	03-May-05	30.75	31.87	28.51	31.3	6819125
AA	01-Jun-05	31.3	33.88	30.41	33.03	4742195
AA	01-Jul-05	32.93	33.25	31.25	32.15	4680407
AIG	02-Jan-05	66.5	70.81	66.38	69.45	5521640
AIG	02-Feb-05	69.6	75.66	69.53	74	5550189

Figure 14-2 The ACTIVITY table (partial data shown)

- Symbol—As previously stated, the Symbol field shows the company's stock exchange symbol.
- Price Date—This field shows the date on which the subsequent prices are taken.
- Open—This is the opening stock price at the start of the day.
- High—This is the stock's highest price during the day.
- Low—This is the stock's lowest price during the day.
- Close—This is the stock's closing price at the end of the day.
- Volume—This is the number of common shares traded during the day.

Figure 14-2 shows all seven records for Alcoa Aluminum (AA) and the first two records for AIG. All 98 records in the table have the same structure, of course.

FINANCIAL Table

The FINANCIAL table, shown in Figure 14-3, shows various financial statistics about each company. A discussion of the table's fields follows the figure.

Financial : Table

Symbol	PE Ratio	Price-to-Book Ratio	Price-to-Sale Ratio	PEG Ratio
AA	19.77	2.2	1.2	1.27
AIG	17.45	2.3	2.1	1.01
AXP	19.8	3.9	2.3	1.4
C	12.06	2.3	3.9	0.92
CAT	19.55	4.1	1.1	1.17
DD	38.74	4.2	1.5	1.93
GE	21.1	3.9	2.4	2.1
HD	16.99	3.3	1.1	1.06
HON	22.5	2.9	1.3	1.79
HPQ	20.32	1.5	0.8	1.34
KO	26.1	8.2	5.6	2.24
MMM	24.06	7.7	3.5	2.05
MO	10.46	3.5	1.1	1.12
XOM	15.02	3.2	1.2	1.85

Figure 14-3 The FINANCIAL table

- PE Ratio—The Price-Earnings Ratio (PE Ratio) is the ratio of the stock's closing price to its earnings per share. *Example*: A company's stock is priced at $100 per share on the exchange. Net income after taxes, divided by the number of shares outstanding on the exchange, is 12. Thus, its PE Ratio would be: 100 / 12 = 8.33.
- Price-to-Book Ratio—This is the ratio of the stock's closing price to its book value per share. The book value per share is the value of the stockholders' equity on its books of account, divided by the number of shares outstanding on the exchange. *Example*: A company's stock is priced at $100 per share on the exchange. Its book value is $10. The price-to-book ratio is 100 / 10 = 10.
- Price-to-Sale Ratio—This ratio is the ratio of the stock's closing price to the amount of sales generated per share of common stock outstanding. *Example*: A company's stock is priced at $100 per share on the exchange. The ratio of its yearly sales to the number of shares outstanding is 20. The price-to-sale ratio is 100 / 20 = 5.
- PEG Ratio—The Price-Earnings Growth Ratio (PEG Ratio) is the ratio of its current PE ratio divided by the expected rate of earnings growth. *Example*: A company's current PE ratio is 16. The company expects its earnings per share to grow 10% a year. The PEG ratio would be: 16 / 10 = 1.6.

Your database is called **STKLOOK.mdb**. To continue, you will need to obtain **STKLOOK.mdb** from your instructor.

ASSIGNMENT 1 USING ACCESS AND EXCEL FOR DECISION SUPPORT

Your uncle knows which stocks he *might* want to buy—these are the stocks in your database. In addition, your uncle has some "rules of thumb" that he wants to follow in making his investments. (These rules of thumb are discussed after Figure 14-4.) He needs a way to experiment with the data—a way to put possible million-dollar portfolios together in a spreadsheet. After he has experimented with data and applied his rules of thumb, he can buy the stocks for the portfolio that looks best to him.

Your spreadsheet should have the following three worksheets. The worksheets, and what to do with each, are discussed next.

- FINANCIAL / NON-FINANCIAL PE RATIOS Worksheet
- ACTIVITY Worksheet
- DATA ABOUT STOCKS Worksheet

FINANCIAL / NON-FINANCIAL PE RATIOS Worksheet

Your uncle wants to know whether the average PE ratios of the financial stocks differ significantly from the average PE ratios of the non-financial companies. Your first step is to import the following data from the database into the FINANCIAL / NON-FINANCIAL PE RATIOS Worksheet: company name, whether it is financial, and the PE ratio. The data should look like that shown in Figure 14-4.

	A	B	C
1	FINANCIAL / NON-FINANCIAL PE RATIOS		
2			
3	COMPANY	FINANCIAL	PE RATIO
4	GE	FALSE	21.10
5	3M	FALSE	24.06

Figure 14-4 FINANCIAL/NON-FINANCIAL PE RATIOS data (partial data shown)

Only partial data is shown in Figure 14-4. In fact, there would be a row of data for each of the 14 companies. This data, alone, is not in a single table, so *you will need to create a select query in Access that collects this data*. (Select queries are discussed in Tutorial B.) You will then import the query's output to the worksheet. (*Note*: Yes/No values imported from Access will be shown as True/False values by Excel. Do not worry about the change in terminology.)

You should then make a pivot table beneath the data that shows the average PE ratio for financial companies, and the average PE ratio of non-financial companies. (Pivot table procedures are discussed in Tutorial E.)

Having done that, you should then enter a formula beneath the pivot table that computes the ratio of the average financial PE ratio to the average non-financial PE ratio. Using that ratio, you should output some guidance for your uncle (which embodies one of his rules of thumb). The rules are summarized in the table in Figure 14-5.

Rule of Thumb	
Average financial PE ratio divided by average non-financial PE ratio	**Action to Take**
< .8	Favor non-financial stocks in portfolio.
> 1.2	Favor financial stocks in portfolio.
Between .8 and 1.2	Have a balance of each kind of stock.

Figure 14-5 PE ratio rules of thumb and actions to take

Example: In the pivot table's cells, the average financial PE ratio is 15, and the average non-financial PE ratio is 25. The ratio is: 15 / 25 = .6. Your uncle would want to buy all, or almost all, non-financial stocks.

ACTIVITY Worksheet

This worksheet shows the closing share price data for each company. Partial data is shown in Figure 14-6.

	A	B	C
1	COMPANY	CLOSE	PRICE DATE
2	3M	79.09	1/2/2005
3	3M	78.02	2/2/2005
4	3M	81.87	3/1/2005
5	3M	86.48	4/1/2005
6	3M	84.56	5/3/2005
7	3M	90.01	6/1/2005
8	3M	85	7/1/2005
9	AIG	69.45	1/2/2005
10	AIG	74	2/2/2005

Figure 14-6 ACTIVITY Worksheet data

There would be 98 rows of data. This selection of data, alone, is not in a single table, so once again *you will need to create a select query in Access that collects this data*. You will import the query's output to the sheet. This data will be useful to your uncle in assessing the trend of prices, as you will later see.

DATA ABOUT STOCKS Worksheet

The data required for this worksheet is revealed in a series of figures. To begin, Figure 14-7 shows some of this data: Company name, Symbol, Financial (whether the company is a financial company), PE ratio, Price-to-book ratio, Price-to-sales ratio, PEG ratio, and the most recent Close price.

	A	B	C	D	E	F	G	H
1	DATA ABOUT STOCKS							
2	COMPANY	SYMBOL	FINANCIAL	PE RATIO	PRICE-TO-BOOK RATIO	PRICE-TO-SALES RATIO	PEG RATIO	CLOSE
3	GE	GE	FALSE	21.10	3.90	2.40	2.10	33.21
4	3M	MMM	FALSE	24.06	7.70	3.50	2.05	85.00

Figure 14-7 DATA ABOUT STOCKS Worksheet (partial data)

This selection of data, alone, is not in a single table. You can get this data in one of two ways: (1) *You can create a select query in Access that collects this data*, then import the query's output to the worksheet; or (2) You can import data from entire Access tables, then delete redundant columns (select the column, select Edit—Delete).

You should use conditional formatting in the Financial column—if a company is a financial company, then the value *TRUE* should appear in bold italics (***TRUE***) and in some color other than the default color

For each of the 14 stocks, you will make five calculations in Excel. This data should be shown to the right of the data in Figure 14-7. The output of those calculations is shown in Figure 14-8. A discussion of the calculations follows the figure.

	I	J	K	L	M
1					
2	PEG OK?	PE OK?	PE BOOK OK?	PE SALES OK?	3-4 OK?
3	NOT OK	NOT OK	OK	NOT OK	NO
4	NOT OK	NOT OK	NOT OK	NOT OK	NO
5	OK	OK	OK	OK	YES

Figure 14-8 Calculations for each stock (partial data shown)

The calculations in Figure 14-8 should be done by =IF() statements:

- PEG OK?—If the PEG ratio is greater than 2, then the stock price is thought to be overvalued, and it would be NOT OK to invest in the stock; otherwise, it would be OK to invest in it.
- PE OK?—If the PE ratio is greater than 20, then the stock price is thought to be overvalued, and it would be NOT OK to invest in the stock; otherwise, it would be OK to invest in it.
- PE BOOK OK?—If the price-to-book ratio is greater than 4, then the stock price is thought to be overvalued, and it would be NOT OK to invest in the stock; otherwise, it would be OK to invest in it.
- PE SALES OK?—If the price-to-sales ratio is less than or equal to 2, then the stock price is thought to be undervalued, and it would be OK to invest in the stock; otherwise, it would be NOT OK to invest in it.
- 3-4 OK?—Looking at the previous four calculations, are there more than two *OK*s in the row for the stock? If so, then a *YES* should show, to indicate that the stock could be purchased; otherwise, a *NO* should show to indicate that it should not be purchased. Here, use the COUNTIF() function to compute the number of *OK*s in the calculation range. If that number exceeds two, your uncle might buy shares of the stock. If it does not exceed two, your uncle would not buy that stock.

At this point, the user would be in a position to construct a portfolio. Data about stocks are included in the portfolio calculation. Then a stock's desirability is checked graphically. These two steps are discussed next.

Portfolio Construction

Beneath the data about the stocks in the DATA ABOUT STOCKS Worksheet, you should have an area where your uncle can play "what-if" to construct the portfolio. The form of this area is shown in Figure 14-9, including some sample data.

	A	B	C	D	E
18	**AVERAGE PE RATIO RECOMMENDATION:**			Strongly favor non-financial	
19					
20	**CONSTRUCT A MILLION DOLLAR PORTFOLIO HERE**				
21	**COMPANY**	**# OF SHARES**	**CLOSE**	**VALUE**	**% OF TOTAL**
22	GE	0	33.21	0	0.00%
23	3M	0	85.00	0	0.00%
24	ALCOA	7500	32.15	241125	24.24%
25	ALTRIA	0	48.77	0	0.00%
26	AMEX	0	49.12	0	0.00%
27	AIG	0	68.78	0	0.00%
28	CATERPILLAR	5000	76.95	384750	38.67%
29	CITIGROUP	0	44.05	0	0.00%
30	COKE	7500	49.20	369000	37.09%
31	DUPONT	0	42.75	0	0.00%
32	EXXON	0	45.17	0	0.00%
33	HP	0	20.00	0	0.00%
34	HOME DEPOT	0	33.99	0	0.00%
35	HONEYWELL	0	35.90	0	0.00%
36					
37	TOTAL			994875	100.00%

Figure 14-9 Form of portfolio-construction area (sample data)

This area should start with a repeat of the Average PE ratio recommendation, which was computed in the FINANCIAL / NON-FINANCIAL PE RATIOS Worksheet. This output will act as guidance in what stocks to select. Use "exclamation point" notation to move the recommendation from a cell in the other sheet to this location. A stock may be put into the portfolio only if it (1) conforms to the Average PE ratio recommendation and (2) it receives a "YES" in the 3-4 OK? column. A portfolio does not have to have *all* the stocks that meet those two restrictions, however. Thus more than one portfolio could be constructed from all the stocks that do satisfy the two restrictions.

Portfolio construction in this area is discussed next.

- COMPANY—These are the company names, which can be copied from the rows in the sheet above.
- # OF SHARES—These are the number of shares that would be bought. The user enters these values manually.
- CLOSE—These are the closing prices for the stocks, which can be copied from the rows in the worksheet above.
- VALUE—This is the value of the shares bought. It is the number of shares multiplied by the closing price.
- % OF TOTAL—This is the percentage of a stock's value to the total amount invested. *Note*: Use absolute addressing properly here.
- TOTAL—This row shows the total invested and the total percentage invested, which should, of course, equal 100%.

Your uncle's rules of thumb and other guidelines should be followed:

1. If the guidance is that non-financial stocks be bought, then the user would enter no amounts in # of Shares for financial stocks (you need not enforce this by rule—the user would merely know to follow it). The user would behave analogously if the guidance was to prefer financial stocks. The user would enter values for some of each kind of stock if the guidance was to balance the portfolio.

2. The user would enter amounts in # of Shares *only* if *YES* shows in the 3-4 OK? column. Again, you do not need formulas to enforce this rule—the user would merely know to follow it.
3. The total invested should not exceed $1 million; the user should enter amounts in the # of Shares column so the total invested comes close to $1 million but does not exceed it.
4. In the example data shown, a portfolio was shown that has three stocks. Your uncle does not have a rule of thumb on the number of stocks. Any number of stocks could be in the portfolio. However, your uncle does not want one company's stock to represent more than 40% of the total value of his investment. Thus, the % of Total column should not have any numbers greater than 40%. (You do not need formulas to enforce this rule—the user would merely know to follow it.)

The portfolio construction area can be used to allocate the million dollars. However, this would not be enough for your uncle. He wants to know that stocks selected show a proper price trend. In his way of thinking, the overall trend should be up, as shown in the price ACTIVITY data for the last seven months. To show that for the stocks in the candidate portfolio, you would move to the ACTIVITY Worksheet and chart the activity for the stocks selected.

Charting Trends

Your first step would be to make a line chart for each of the candidate stocks in the portfolio. An example is shown in Figure 14-10, continuing the example portfolio.

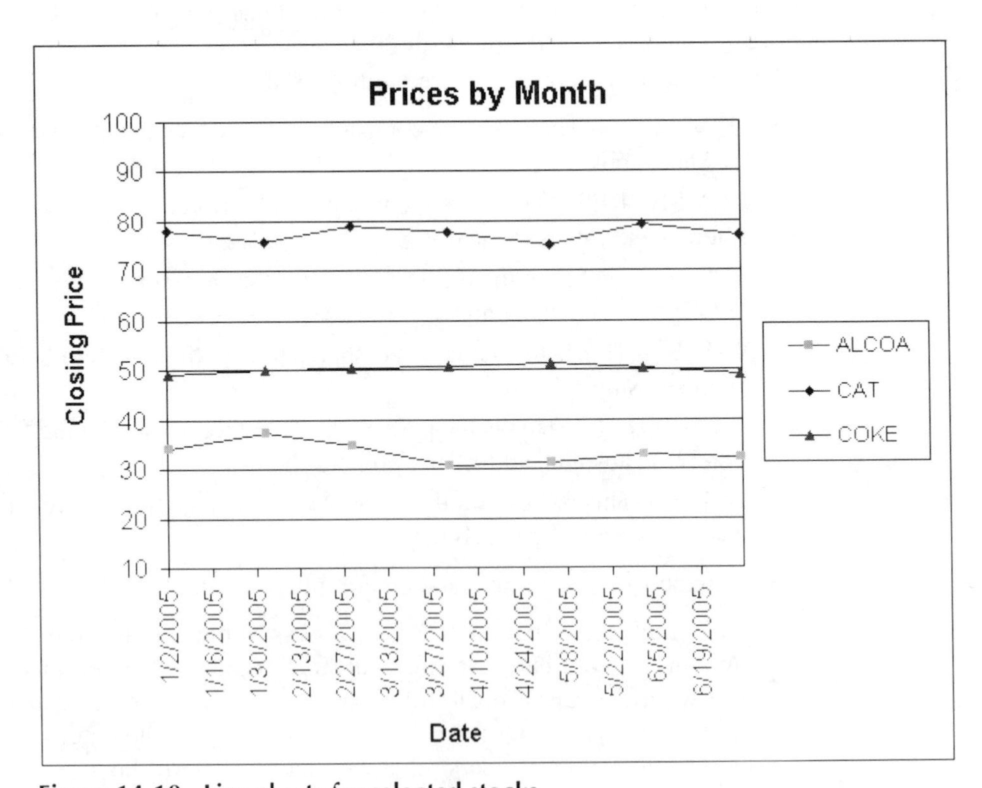

Figure 14-10 Line charts for selected stocks

Conceivably, the trend of prices will be obvious by inspection, but to be sure, a trend line should be inserted. Your next step is to place a trend line (called a *trendline* in Excel) on each line graph. Follow these steps:

1. Click on the line graph for the stock's activity.
2. Then select **Chart—Add Trendline**.
3. Choose a **Linear** trendline, and then click the **OK** button. The trendline will show up on top of the line graph.

Continuing the example, the results are shown in Figure 14-11.

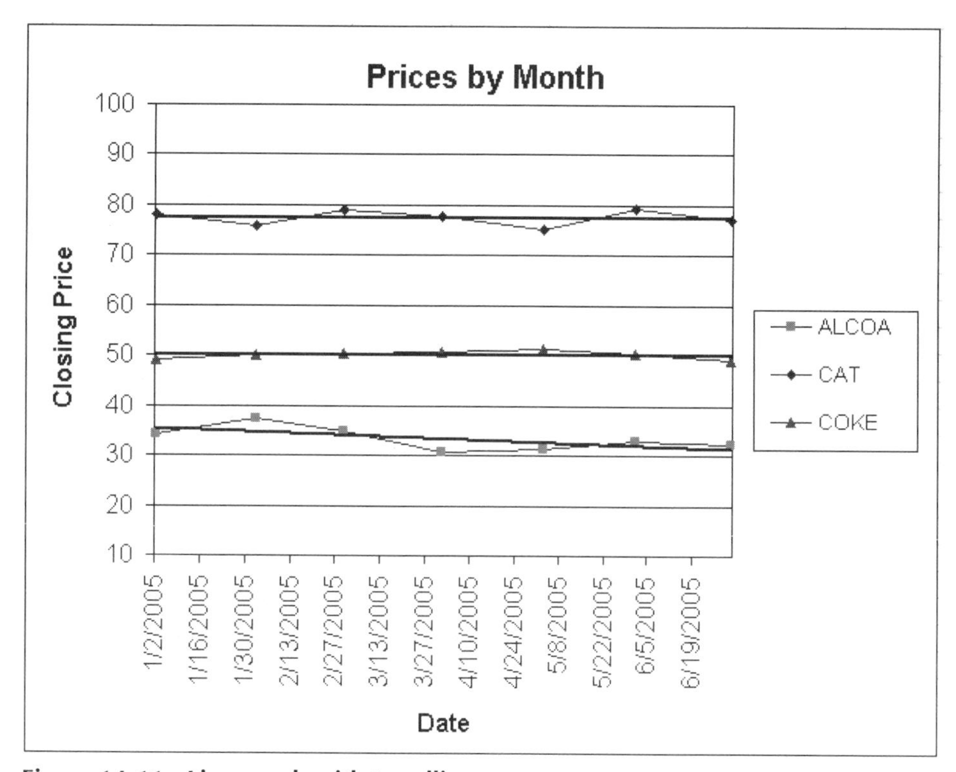

Figure 14-11 Line graph with trendlines

Again, the trends for each stock might be obvious by inspection. To be more precise, however, the line's formula should be shown. A straight line has this formula:

$$Y = mx + b$$

The *Y* value is a function of the *x* value. In this case, *Y* is the price, and *x* is time (month). Here, Excel sees the January value as *1*, the February value as *2*, and so forth. The *m* represents the slope of the line. Upward trends are indicated by a positive value for *m*. Your uncle wants all stock trendlines to have a positive slope!

To insert the trendline formula do the following:

1. Right-click on the **Trendline**.
2. Select **Format Trendline** in the menu.
3. Click the **Options** tab.
4. Select **Display equation** on chart, and then select **OK**. The formula will display, but you may need to click and drag the formula to size it properly for viewing.

Case 14

Continuing the example, doing this for each trendline would reveal the formula, as shown in Figure 14-12:

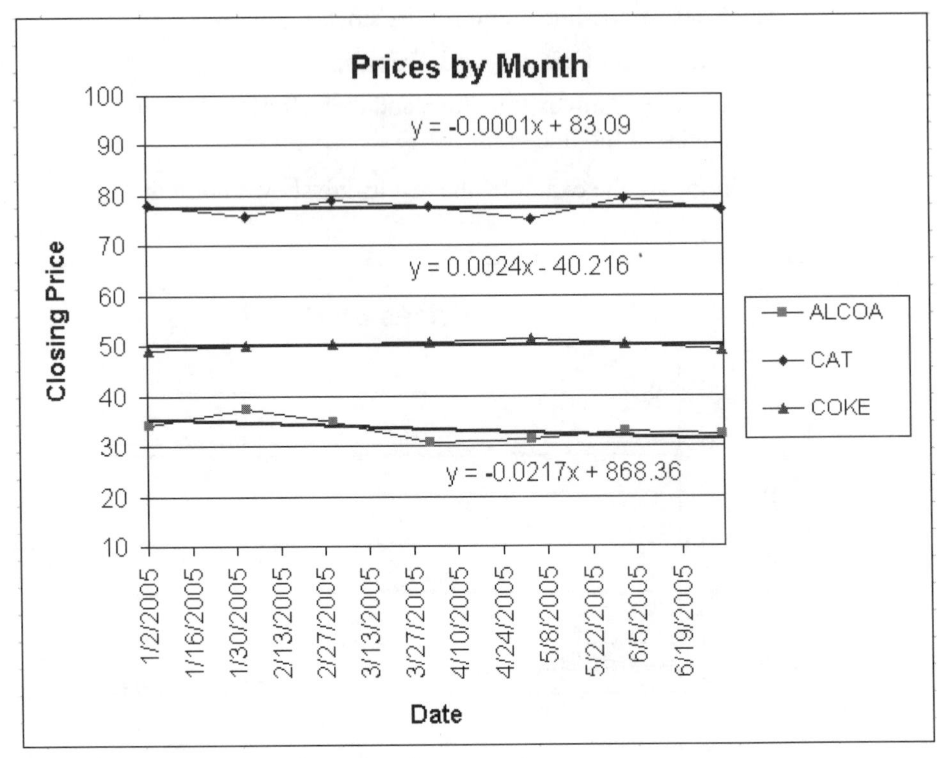

Figure 14-12 Trendline formulas shown in chart

Here, you will note that two of the trendlines have a negative slope, so the associated stocks would not be acceptable for your uncle. Your uncle would have to go back to the construction area and replace those stocks with others.

ASSIGNMENT 2 USING THE SPREADSHEET FOR DECISION SUPPORT

You should use Access and Excel to construct a portfolio for your uncle's consideration, following the procedures described previously.

When you are finished with the spreadsheet, follow these steps:

1. Save the file one last time (**File—Save**). A good file name would be PORTFOL.xls.
2. Then, with the disk in **Drive A**:, select **File—Close** and then **File—Exit**.

You are now in a position to document your work in a memorandum. Write a memorandum to your uncle about your work. Observe the following requirements:

- Your memorandum should have a proper heading (DATE / TO / FROM / SUBJECT). You may wish to use a Word memo template (**File**—click **New**, click **On my computer** in the Templates section, click the **Memos** tab, choose **Contemporary Memo**, then click **OK**).
- Briefly outline the situation and your portfolio construction. You may wish to note that your uncle can continue to play "what if" with the variables.

- Support your presentation graphically in two ways: (1) by inserting a summary table after the prose, similar to that shown in Figure 14-13; (2) by attaching printouts of your trendline charts for the portfolio you have created.

Company	Number of Shares	Price	Value	% of Total
Company 1				
Company 2				
...				
Total				100%

Figure 14-13 Form for table in memorandum

Enter a table into Word, using the following procedure:

1. Select the **Table** menu option, point to **Insert**, then click **Table**.
2. Enter the number of rows and columns.
3. Select **AutoFormat** and choose **Table Grid 1.**
4. Select **OK**, and then select **OK** again.

Assignment 3 Giving an Oral Presentation

Assume that your uncle is impressed by your "what-if" portfolio program and by your analysis. He is a member of an investment club and thinks the club's members might want to use the program for their own investing! Maybe you can sell the program to each club member! He asks you to give a presentation explaining your program and your method to his club's members. Bear in mind that audience members are not software experts. Prepare to explain your work and your recommendation to the group in 10 minutes or fewer. Use visual aids or handouts that you think are appropriate. Tutorial E has guidance on how to prepare and give an oral presentation.

Deliverables

1. Printout of your memorandum, including printouts of charts
2. Spreadsheet printouts
3. Disk or CD, which should have your Word memorandum file, your Excel spreadsheet file, and your Access .mdb file.

Staple the printouts together, with the memorandum on top. If there are other .xls files or .mdb files on your disk or CD, write your instructor a note, stating the names of the files pertinent to this case.

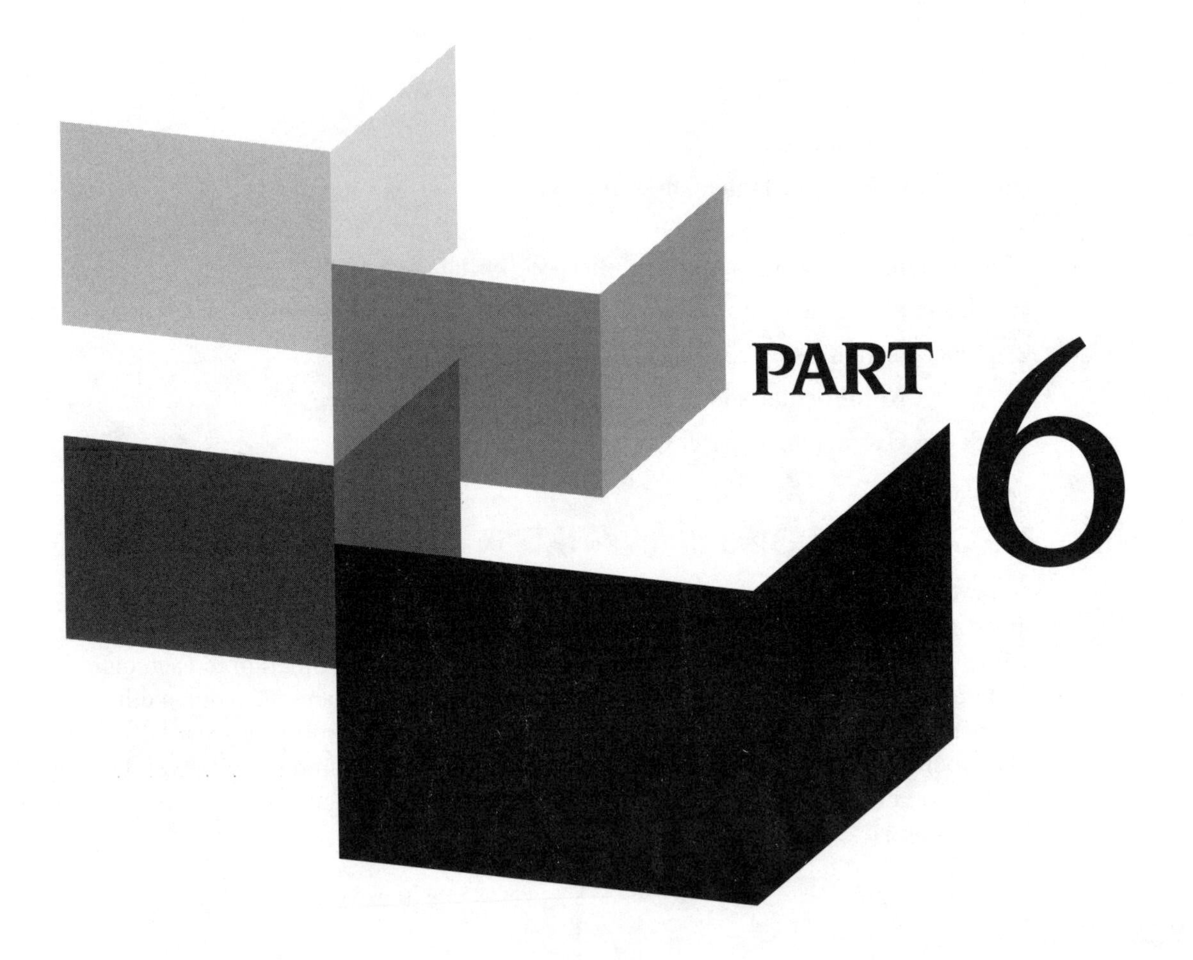

Presentation Skills

Giving an Oral Presentation

TUTORIAL

Giving an oral presentation provides you with the opportunity to practice the presentation skills you'll need in the workplace. The presentations you'll create for the cases in this textbook will be similar to real-world presentations: You'll present objective, technical results to an organization's stakeholders. You'll support your presentation with visual aids commonly used in the business world. Your instructor may wish to have your classmates role-play an audience of business managers, bankers, or employees and have them give you feedback on your presentation.

Follow these four steps to create an effective presentation:

1. Plan your presentation.
2. Draft your presentation.
3. Create graphics and other visual aids.
4. Practice your delivery.

Let's start at the beginning and look at the steps involved in planning your presentation.

Plan Your Presentation

When planning an oral presentation, you'll need to know your time limits, establish your purpose, analyze your audience, and gather information. Let's look at each of these elements.

Know Your Time Limits

You'll need to consider your time limits on two levels. First, consider how much time you'll have to deliver your presentation. What can you expect to accomplish in 10 minutes? The element of time is the "driver" of any presentation. It limits the breadth and depth of your talk—and the number of visual aids that you can use. Second, consider how much time you'll need for the actual process of preparing your presentation: drafting your presentation, creating graphics, and practicing your delivery.

Establish Your Purpose

You must define your purpose: what you need and want to say and to whom. For the cases in the Access portion of the book, your purpose will be to inform and explain. For instance, a business's owners, managers, and employees need to know how their organization's database is organized and how to use it to fill in input forms, create reports, and so on. By contrast, for the cases in the Excel portion of the book, your purpose will be to recommend a course of action. You'll be making recommendations based on your results from inputting various scenarios to business owners, managers, and bankers.

Analyze Your Audience

Before drafting your presentation, analyze your audience. Ask yourself these questions: What does my audience already know about the subject? What do they want to know? What do they need to know? Do they have any biases that I should consider? What level of technical detail is best suited to their level of knowledge and interest?

In some Access cases, you will make a presentation to an audience who might not be familiar with Access or databases in general. In other cases, you might be giving a presentation to a business owner who started work on the database but was not able to finish it. Tailor your presentation to suit your audience.

For the Excel cases, you will be interpreting results for an audience of bankers and business managers. The audience does not need to know the detailed technical aspects of how you generated your results. What they *do* need to know is what assumptions you made prior to developing your spreadsheet because those assumptions might have an impact on their opinion of your results.

Gather Information

Since you will have just completed a case, you'll have the basic information. For the Access cases, review the main points of the case and your goals. Be sure to include all the points that you feel are important for the audience. In addition, you may wish to go beyond the requirements and explain additional ways in which the database could be used to benefit the organization, now or in the future.

For the Excel cases, you can refer to the tutorials for assistance in interpreting the results from your spreadsheet analysis. For some cases, you might want to research the Internet for business trends or background information that could be used to support your presentation. For example, for Case 14, Your Uncle's Portfolio Contruction Problem, you may want to do Internet research so you can more confidently predict the future of the economy and the financial marketplace.

DRAFTING YOUR PRESENTATION

You might be tempted to write your presentation and then memorize it, word for word. If you do, your presentation will sound very unnatural because when people speak, they use a simpler vocabulary and shorter sentences than when they write. Thus, you may want to draft your presentation by noting just key phrases and statistics. When drafting your presentation, follow this sequence:

1. Write the main body of your presentation.
2. Write the introduction to your presentation.
3. Write the conclusion to your presentation.

Writing the Main Body

When you draft your presentation, write the body first. If you try to write the opening paragraph first, you'll spend an inordinate amount of time creating "the perfect paragraph"—only to revise it after you've written the body of your presentation.

Keep Your Audience in Mind

To write the main body, review your purpose and your audience's profile. What are the main points you need to make? What are your audience's wants, needs, interests, and technical expertise? It's important to include some basic technical details in your presentation, but keep in mind the technical expertise of your audience.

What if your audience consists of people with different needs, interests, and levels of technical expertise? For example, in the Access cases, an employee might want to know how to input information into a form, but the business owner might already know how to input data and may be more interested in generating queries and reports. You'll need to acknowledge their differences in your presentation. Thus, you might want to say something like, "And now, let's look at how data entry clerks can input data into the form."

Similarly, in the Excel cases, your audience will usually consist of business owners, managers, and bankers. The owners' and managers' concerns will be profitability and growth. By contrast, the bankers' main concern will be getting a loan repaid. You'll need to address the interests of each group.

Use Transitions and Repetition

Because your audience can't read the text of your presentation, you'll need to use transitions to compensate. Words such as *next*, *first*, *second*, *finally*, etc., will help your audience follow the sequence of your ideas. Words such as *however*, *by contrast*, *on the other hand*, and *similarly* will help them to follow shifts in thought. You can also use your voice and hand gestures to convey emphasis.

Also think about how you can use body language to emphasize what you're saying. For instance, if you are stating three reasons, you can tick them off on your fingers as you discuss them: one, two, three. Similarly, if you're saying that profits will be flat, you can make a level motion with your hand for emphasis.

As you draft your presentation, repeat key points to emphasize them. For example, suppose that your point is that outsourcing labor will provide the greatest gains in net income. Begin by previewing that concept: State that you're going to demonstrate how outsourcing labor will yield the greatest profits. Then provide statistics that support your claim and show visual aids that graphically illustrate your point. Summarize by repeating your point: "As you can see, outsourcing labor does yield the greatest profits."

Rely on Graphics to Support Your Talk

As you write the main body, think of how you can best incorporate graphics into your presentation. Don't waste a lot of words describing what you're presenting if you can use a graphic that can quickly portray it. For instance, instead of describing how information from a query is input into a report, show a sample, a query result, and a completed report. Figures E-1 and E-2 illustrate this.

Query for Report : Select Query

Last Name	First Name	Date	Type	Amount Paid
Angerstein	Amy	6/30/2006	Catalina 23	20
Angerstein	Amy	7/2/2006	Catalina 23	35
Codfish	Colburn	7/5/2006	Capri 14.2	60
Dellabella	Debra	7/5/2006	Hobie 18	20
Edelweiss	Edna	7/11/2006	Sunfish 12	150
Angerstein	Amy	8/1/2006	Row	30

Figure E-1 Access query

Summary Report

Last Name	*First Name*	*Date*	*Type*	*Amount Paid*
Angerstein	*Amy*			
		8/1/2006	Row	$30.00
		7/2/2006	Catalina 23	$35.00
		6/30/2006	Catalina 23	$20.00
	Total			*$85.00*
Codfish	*Colburn*			
		7/5/2006	Capri 14.2	$60.00
	Total			*$60.00*

Figure E-2 Access report

Also consider what kinds of graphics media are available—and how well you know how to use them. For example, if you've never prepared a PowerPoint presentation, will you have enough time to learn how to do it before your presentation?

Anticipate the Unexpected

Even though you're just drafting your report now, eventually you'll be answering audience questions. Being able to handle questions smoothly is the mark of a professional. The first step is being prepared for those questions.

You won't use all the facts you gather in your presentation. However, as you draft your presentation, you might want to keep some of those facts jotted on paper—just in case you need them to answer questions from the audience. For instance, for some Excel presentations you might be asked why you are not recommending some course of action that you did not mention in your report.

Writing the Introduction

After you have written the main body of your talk, then develop an introduction. An introduction should be only a paragraph or two in length and preview the main points that your presentation will cover.

For some of the Access cases, you might want to include some general information about databases: what they can do, why they are used, and how they can help the company become more efficient and profitable. You won't need to say much about the business operation since the audience already works for the company.

For the Excel cases, you might want to have an introduction to the general business scenario and describe any assumptions you made when creating and running your decision support spreadsheet. Excel is used for decision support, so describe the choices and decision criteria.

Writing the Conclusion

Every good presentation needs a good ending. Don't leave the audience hanging! Your conclusion should be brief—only a paragraph or two in length—and give your presentation a sense of closure. Use the conclusion to repeat your main points or, for the Excel cases, your findings and/or recommendations.

CREATING GRAPHICS

Using visual aids is a powerful method of getting your point across and making it understandable to your audience. Visual aids come in a variety of physical forms. Some forms are more effective than others.

Choosing Graphics Media

The media that you use should depend on your situation and what media are available. One of the key things to remember when using any media is this: *You must maintain control of the media or you'll lose control of your audience.*

The following list highlights some of the most common media and their strengths and weaknesses.

- **Handouts:** This medium is readily available in both classrooms and businesses. It relieves the audience from taking notes. Graphics can be in full color, of professional quality, and multi-colored. *Negatives*: You must stop and take time to hand out individual sheets. During your presentation, the audience might be studying and discussing your handouts rather than listening to you. Lack of media control is *the* major drawback—and it can kill your presentation.
- **Chalkboard:** This informal medium is readily available in the classroom but not in many businesses. *Negatives*: You'll need to turn your back on the audience when you're writing (thus losing control of them), and you'll need to erase what you've written as you go. Your handwriting must be very good. In addition, attractive graphics are difficult to create.
- **Flip Chart:** This informal medium is readily available in many businesses. *Negatives*: The writing space is so small, it's not effective for more than a very small group. This medium shares many of the same negatives as the chalkboard.
- **Overheads:** This medium is readily available in both classrooms and businesses. You do have control over what the audience sees and when. You can create very professional PowerPoint presentations on overhead transparencies. *Negatives*: Handwritten overheads look amateurish. Without special equipment and preparation, graphics are difficult to do well.
- **Slides:** This formal medium is readily available in many businesses and can be used in large rooms. Slides can be either 35mm slides or the more popular electronic on-screen slides, which is usually *the* medium of choice for most large organizations. It is slick and professional and is generally preferred for formal presentations.

Tutorial E

Negatives: You must have access to the equipment and know how to use it. It takes time to learn how to create and use computer graphics. Also, you must have some source of ambient light, or it will be difficult to see your notes in the dark.

Creating Charts and Graphs

Technically, charts and graphs are not the same thing, although many graphs are called "charts." Usually, charts show relationships, and graphs show change. However, Excel makes no distinction and calls both charts.

Charts are easy to create in Excel. Unfortunately, they are so easy to create that people often create graphics that are meaningless or that inaccurately reflect the data they represent. Let's look at how to select the most appropriate graphics.

Charts

Use pie charts to display data that is related to a whole. Excel takes the numbers you want to graph and makes them a percentage of 100. You might use a pie chart when showing the percentage of shoppers who bought a generic brand of toothpaste versus a major brand, as shown in Figure E-3. You would *not*, however, use a pie chart to show a company's net income over a three-year period, because the period cannot be considered "a whole" or the years its "parts," as shown in Figure E-4.

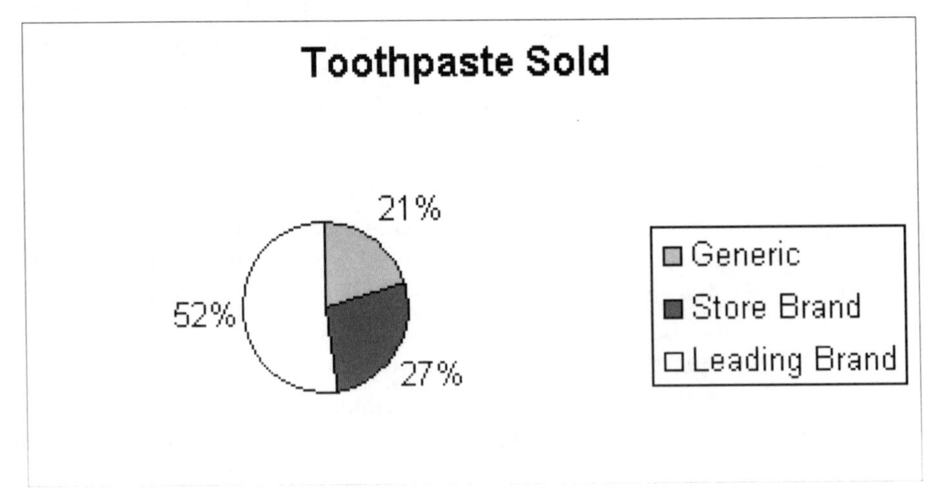

Figure E-3 Pie chart: appropriate use

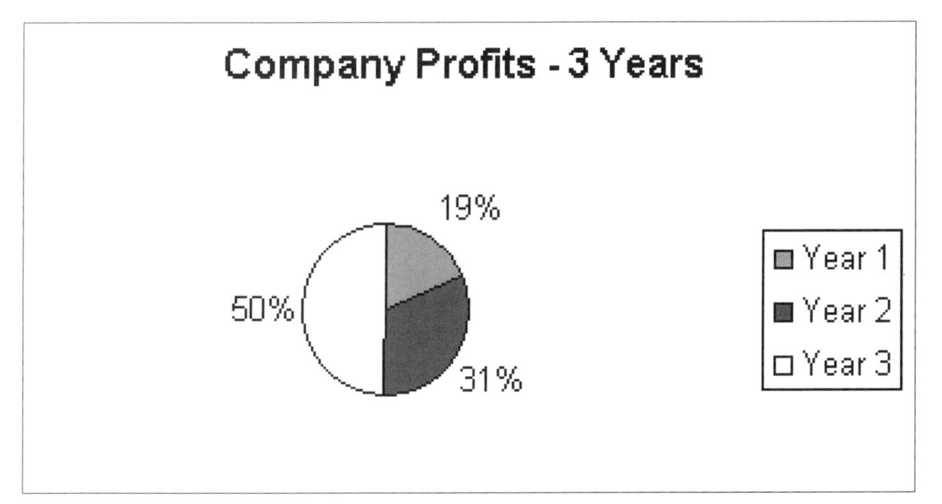

Figure E-4 Pie chart: inappropriate use

Use bar charts when you want to compare several amounts at one time. For example, you might want to compare the net profit that would result from each of several different strategies. You can also use a bar chart to show changes over time. For example, you might show how one pricing strategy would increase profits year after year.

When you are showing a graphic, don't forget that you need labels that explain what the graphic shows. For instance, if you're showing a graph with an X and Y axis, you should show what each axis represents so the audience doesn't puzzle over the graphic while you're speaking. Figures E-5 and E-6 show the necessity of labels.

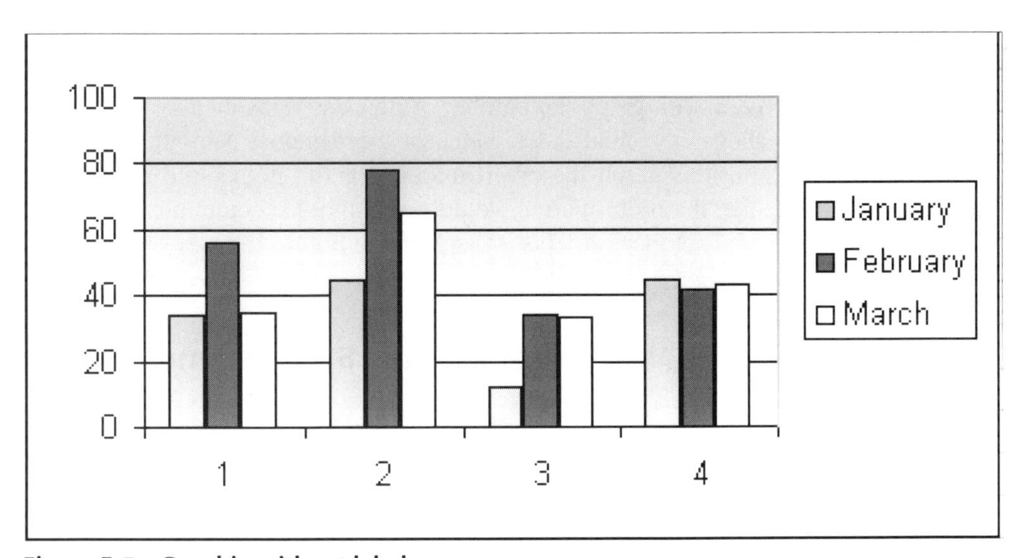

Figure E-5 Graphic without labels

Tutorial E

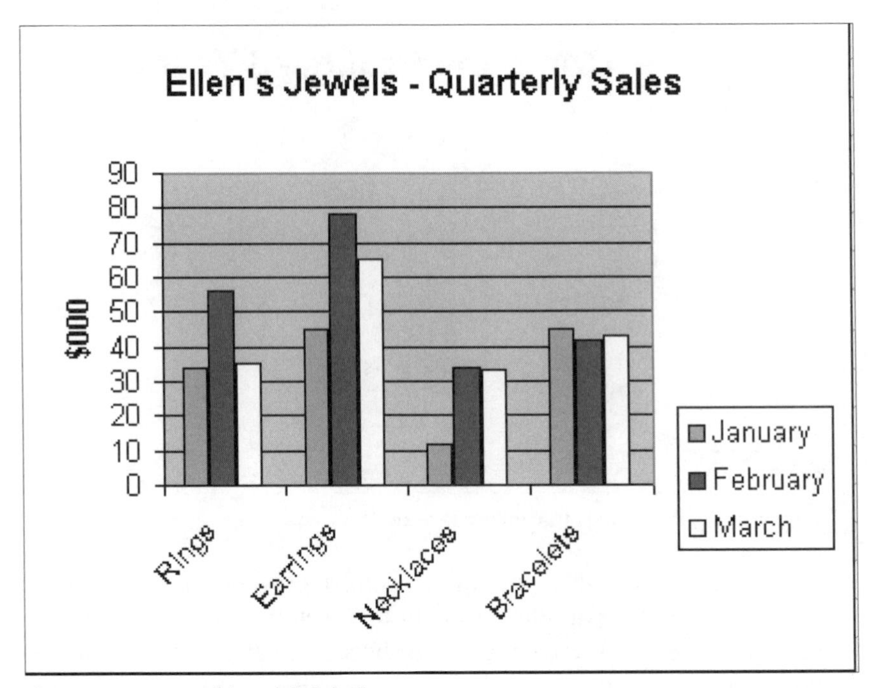

Figure E-6 Graphic with labels

In Figure E-5, the graphic is not labeled, and neither are the X and Y axes: Are the amounts shown units or dollars? What elements are represented by each bar? By contrast, Figure E-6 provides a comprehensive snapshot of the business operation—which would support a talk rather than distract from it.

Another common pitfall is creating charts that have a misleading premise. For example, suppose that you want to show how sales have increased and contributed to a growth in net income. If you graph the number of items sold, as displayed in Figure E-7, it might not tell you about the actual dollar value of those items; it might be more appropriate (and more revealing) to graph the profit margin for the items sold times the number of items sold. Graphing the profit margin would give a more accurate picture of what is contributing to the increased net income. This is displayed in Figure E-8.

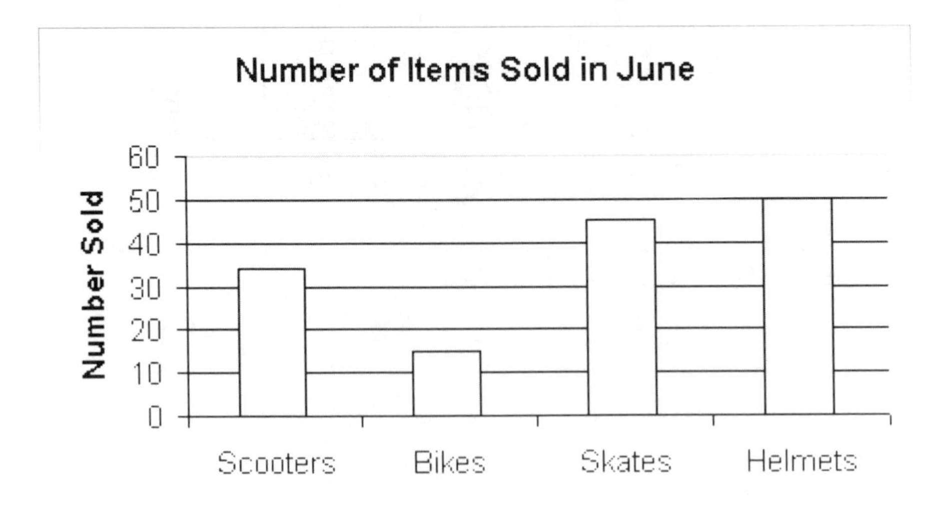

Figure E-7 Graph: number of items sold

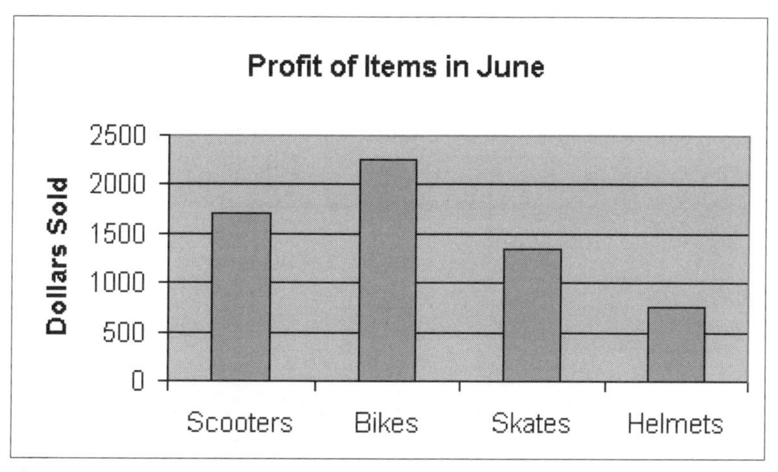

Figure E-8 Graph: profit of items sold

Another common pitfall is putting too much data in a single, comparative chart. Here is an example: Assume that you want to compare monthly mortgage payments for two loans with different interest rates and timeframes. You have a spreadsheet that computes the payment data, shown in Figure E-9.

Calculation of Monthly Payment						
Rate	6.00%	6.10%	6.20%	6.30%	6.40%	6.50%
Amount	100000	100000	100000	100000	100000	100000
Payment (360 payments)	$599	$605	$612	$618	$625	$632
Payment (180 payments)	$843	$849	$854	$860	$865	$871
Amount	150000	150000	150000	150000	150000	150000
Payment (360 payments)	$899	$908	$918	$928	$938	$948
Payment (180 payments)	$1,265	$1,273	$1,282	$1,290	$1,298	$1,306

Figure E-9 Calculation of monthly payment

You try to capture all this information in a single Excel chart, such as the one shown in Figure E-10.

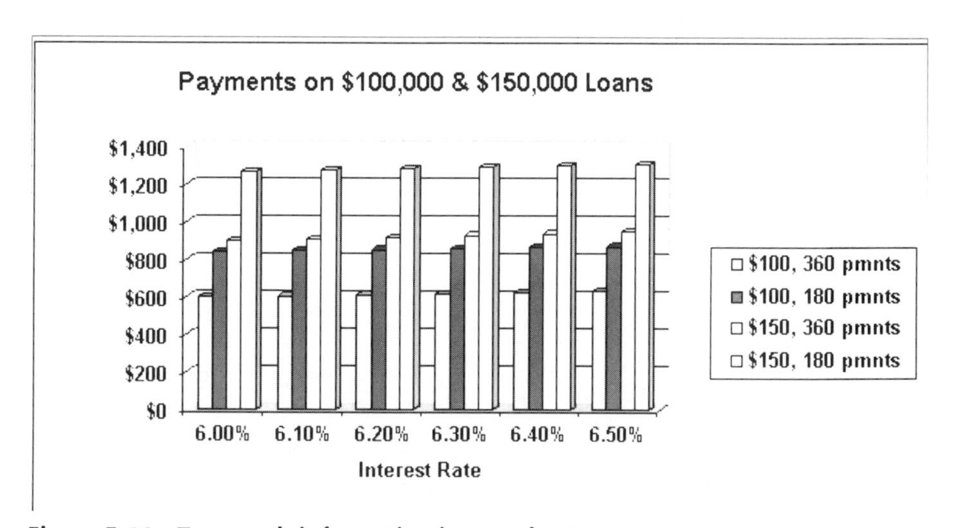

Figure E-10 Too much information in one chart

Tutorial E

There is a great deal of information here. Most readers would probably appreciate it if you broke things up a bit. It would probably be easier to understand the data if you made one chart for the $100,000 loan and another one for the $150,000 loan. The chart for the $100,000 loan would look like the chart shown in Figure E-11.

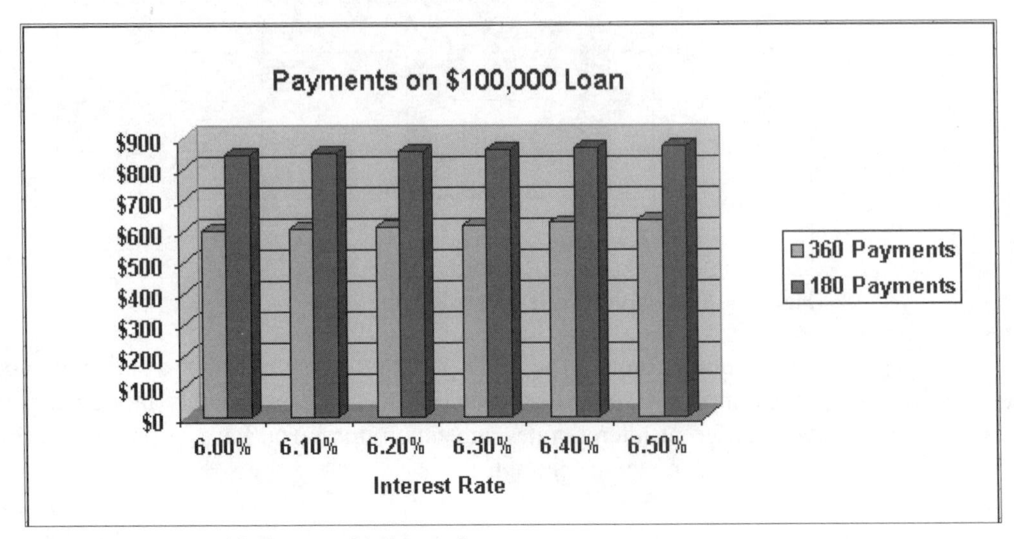

Figure E-11 Good balance of information

A similar chart could be made for the $150,000 loan. The charts could then be augmented by text that summarizes the main differences between the payments for each loan. In this fashion, the reader is led step by step through the data analysis.

You may wish to use the Chart Wizard in Excel, but be aware that the Charting functions can be tricky to use at times, especially with sophisticated charting. Some tweaking to the chart is often necessary. Your instructor may be able to provide specific directions for your individual charts.

Making a PivotTable in Excel

Suppose that you have data for a company's sales transactions by month, by salesperson, and by amount for each product type. You would like to display each salesperson's total sales, according to type of product sold and also by month. Using a PivotTable in Excel, you can tabulate such summary data, using one or more dimensions.

Figure E-12 shows total sales, cross-tabulated by salesperson and by month. This display was created by using a PivotTable in Excel.

	A	B	C	D	E
1	**Name**	**Product**	**January**	**February**	**March**
2	Jones	Product 1	30,000	35,000	40,000
3	Jones	Product 2	33,000	34,000	45,000
4	Jones	Product 3	24,000	30,000	42,000
5	Smith	Product 1	40,000	38,000	36,000
6	Smith	Product 2	41,000	37,000	38,000
7	Smith	Product 3	39,000	50,000	33,000
8	Bonds	Product 1	25,000	26,000	25,000
9	Bonds	Product 2	22,000	25,000	24,000
10	Bonds	Product 3	19,000	20,000	19,000
11	Ruth	Product 1	44,000	42,000	33,000
12	Ruth	Product 2	45,000	40,000	30,000
13	Ruth	Product 3	50,000	52,000	35,000

Figure E-12 Excel spreadsheet data

You can create this kind of table (and many other kinds) with Excel's PivotTable tool. You can use the following steps to create a PivotTable.

1. Select Data—PivotTable and PivotChart Report. You will see the screen shown in Figure E-13.

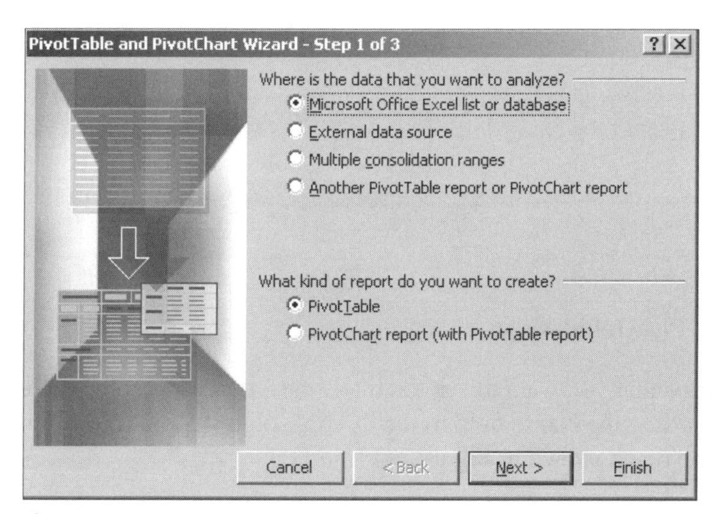

Figure E-13 Step 1

2. To make a PivotTable, click Next. You will see the screen shown in Figure E-14. By default the range will be the most northeast contiguous data range in the spreadsheet. You can change this in the Range window.

Figure E-14 Step 2

3. Click Next. You will then see the screen shown in Figure E-15.

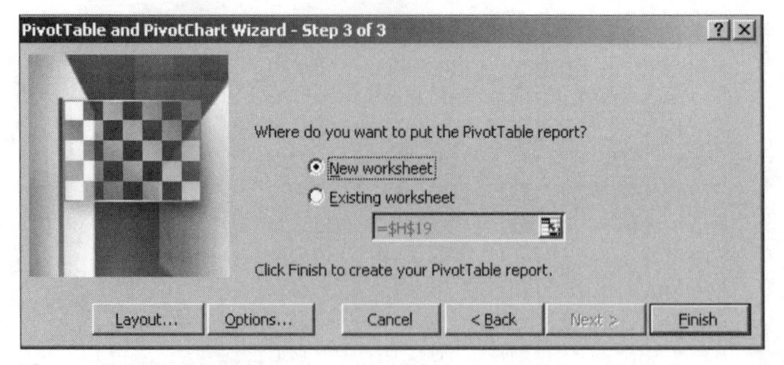

Figure E-15 Step 3

4. You can put the table in the current sheet (probably "Sheet1") or in a separate sheet. The latter way is shown here. *New worksheet* is the default. Click Finish.
5. You will see the screen shown in Figure E-16. The data range's column headings are shown in the PivotTable Field List. Click and drag column headings into the Row, Column, and Data areas.

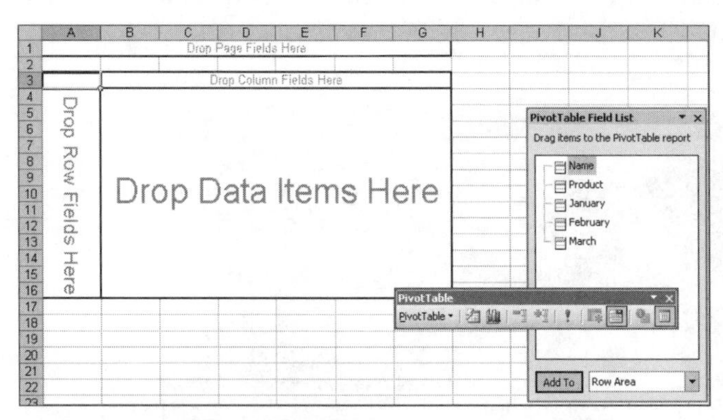

Figure E-16 PivotTable design

6. Assume that you want to see the total sales, by product, for each salesperson. You would drag the Name field to the "Drop Column Fields Here" area, and you should see the result shown in Figure E-17.

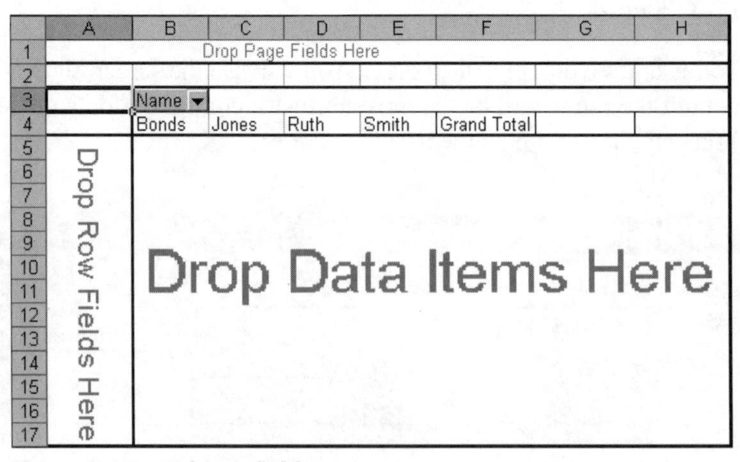

Figure E-17 Column fields

7. Next, take the Product field and drag it to the “Drop Row Fields Here” area, and you should see the result shown in Figure E-18.

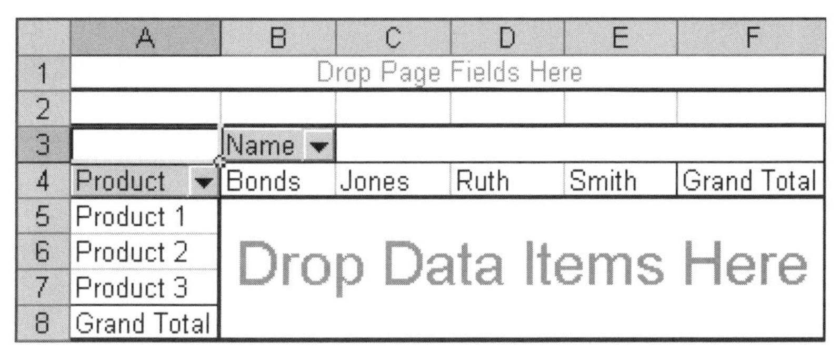

	A	B	C	D	E	F
1	Drop Page Fields Here					
2						
3		Name				
4	Product	Bonds	Jones	Ruth	Smith	Grand Total
5	Product 1	Drop Data Items Here				
6	Product 2					
7	Product 3					
8	Grand Total					

Figure E-18 Row fields

8. Finally, take the month fields (January, February, and March) and drag them individually to the “Drop Data Items Here” area to produce the final PivotTable; you should see the result shown in Figure E-19.

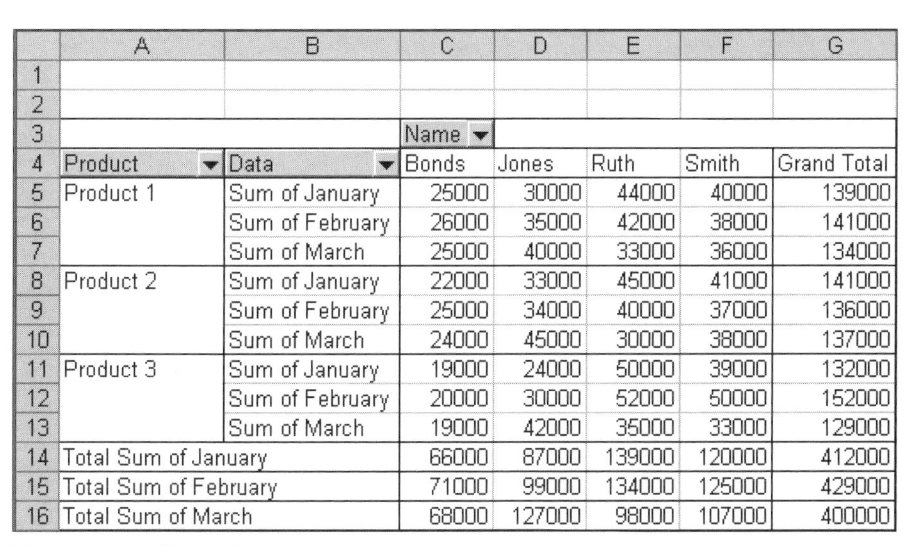

	A	B	C	D	E	F	G
1							
2							
3			Name				
4	Product	Data	Bonds	Jones	Ruth	Smith	Grand Total
5	Product 1	Sum of January	25000	30000	44000	40000	139000
6		Sum of February	26000	35000	42000	38000	141000
7		Sum of March	25000	40000	33000	36000	134000
8	Product 2	Sum of January	22000	33000	45000	41000	141000
9		Sum of February	25000	34000	40000	37000	136000
10		Sum of March	24000	45000	30000	38000	137000
11	Product 3	Sum of January	19000	24000	50000	39000	132000
12		Sum of February	20000	30000	52000	50000	152000
13		Sum of March	19000	42000	35000	33000	129000
14	Total Sum of January		66000	87000	139000	120000	412000
15	Total Sum of February		71000	99000	134000	125000	429000
16	Total Sum of March		68000	127000	98000	107000	400000

Figure E-19 Data items

By default, Excel adds up all the sales for each salesperson by month for each individual product. It also shows the total sales for each month for all products at the bottom of the PivotTable.

Creating PowerPoint Presentations

PowerPoint presentations are easy to create: Simply open up the application and use the appropriate slide layout for a title slide, a slide containing a bulleted list, a picture, a graphic, and so on. In choosing a design template (the background color, the font color and size, and the fill-in colors for all slides in your presentation), keep these guidelines in mind:

- Avoid using pastel background colors. Dark backgrounds such as blue, black, and purple work well on overhead projection systems.

- If your projection area is small or your audience is large, you might want to use boldface type for all your text to make it even more visible.
- Try using "transition" slides to keep your talk lively. A variety of styles are in the program and available for use. Common transitions include "dissolves" and "wipes." Avoid wild transitions, such as swirling letters, that will distract your audience from your presentation.
- You can use "build" effects if you do not want your audience to see the whole slide when you show it. A "build" effect will allow each bullet to come up when the mouse button or the right arrow is clicked. A "build" effect allows you to control the visual and explain the elements as you go. This can be controlled under the Custom Animation screen, as shown in Figure E-20.

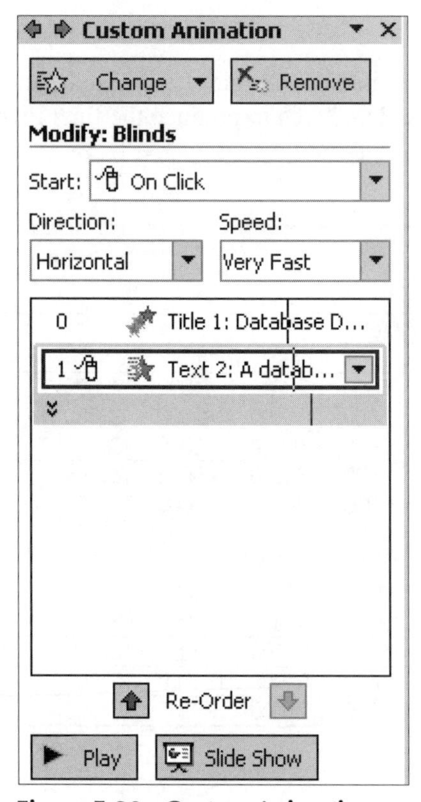

Figure E-20 Custom Animation screen

- You can create PowerPoint slides that have a section for notes. These are printed for the speaker when you choose Notes Pages from the *Print what* drop-down menu on the Print dialog box, as shown in Figure E-21. Each slide is printed as half-size, with the notes written underneath each slide, as shown in Figure E-22.

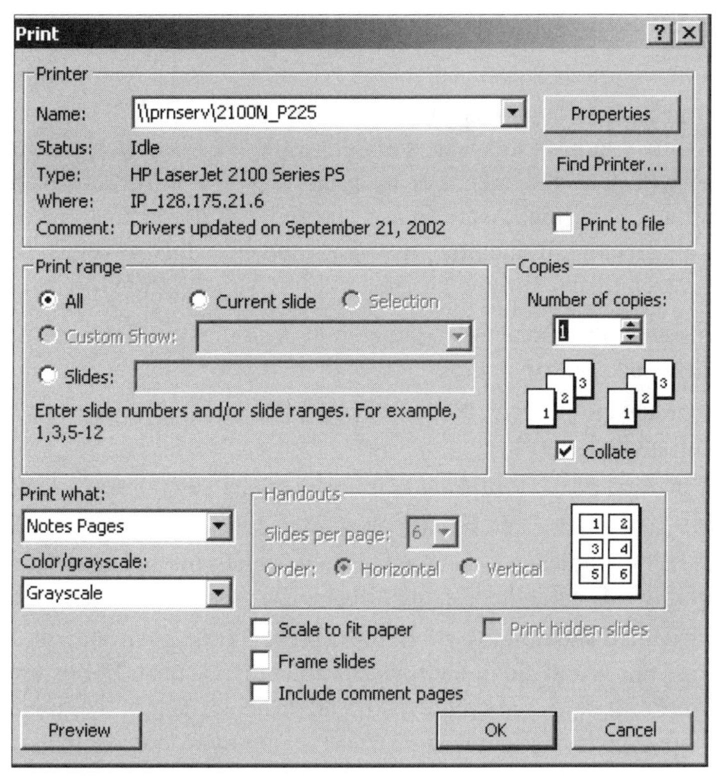

Figure E-21 Printing notes page

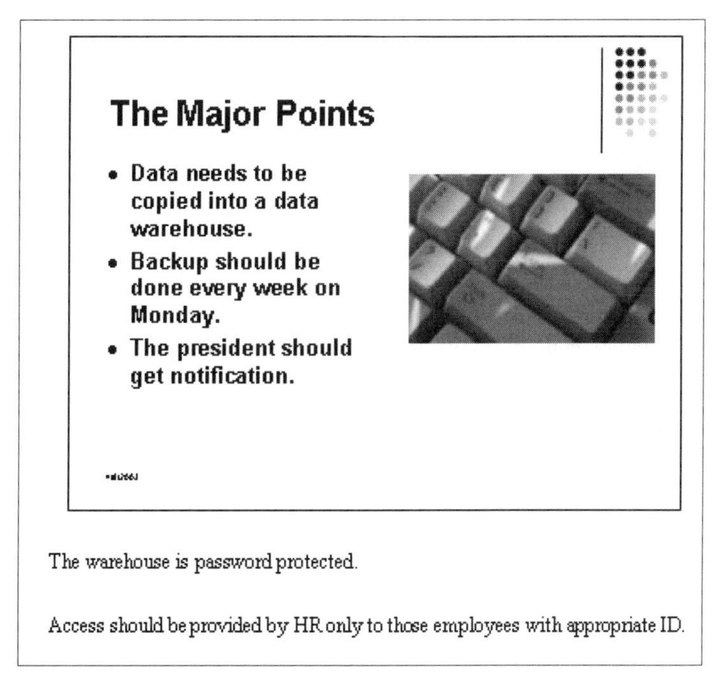

Figure E-22 Sample notes page

- As previously mentioned, always check your presentation on the overhead. What looks good on your computer screen might not be readable on an overhead screen.

Using Visual Aids Effectively

Make sure that you've chosen visual aids that will work for you most effectively. Also make sure that you have enough—but not too many—visual aids. How many is too many? The amount of time you have to speak will determine the number of visual aids that you should use, and so will your audience. For example, if you will be addressing a group of teenage summer helpers, you might want to use more visual effects than if you make a presentation to a board of directors. Remember, use visual aids to enhance your talk, not replace it.

Review each visual aid you've created to make sure that it meets the following criteria:

- The size of the visual aid is large enough so that everyone in the audience can see it clearly and read any labels.
- The visual aid is accurate, e.g., the graphics are not misleading and there are no typos or misspelled words.
- The content of the visual aid is relevant to the key points of your presentation.
- The visual aid doesn't distract the audience from your message. Often when creating PowerPoint slides, speakers get carried away with the visual effects, e.g., they use spiraling text and other jarring effects. Keep it professional.
- A visual aid should look good in the presentation environment. If at all possible, try using your visual aid in the presentation environment. For example, when using PowerPoint, try it out on the overhead projector and in the room in which you'll be showing the slides. What looks good on your computer screen might not look good on the overhead projector when viewed from a distance of 20 feet.
- Make sure that all numbers are rounded unless decimals or pennies are crucial.
- Do not make your slides too busy or crowded. Most experts say that bulleted lists should contain no more than four or five lines. Also avoid having too many labels. A busy slide is illustrated in Figure E-23.

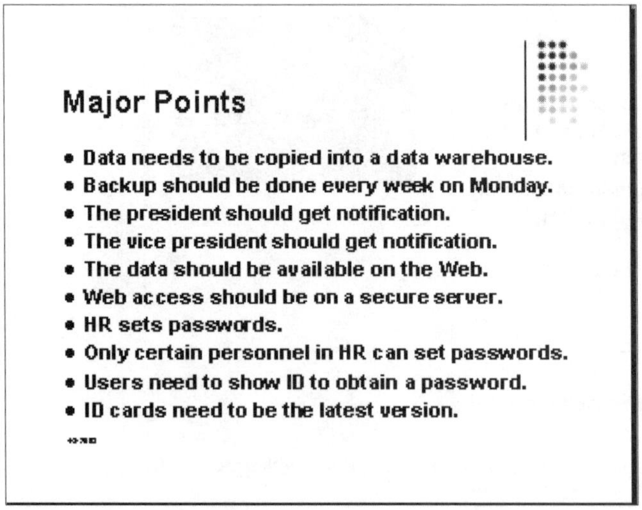

Major Points

- Data needs to be copied into a data warehouse.
- Backup should be done every week on Monday.
- The president should get notification.
- The vice president should get notification.
- The data should be available on the Web.
- Web access should be on a secure server.
- HR sets passwords.
- Only certain personnel in HR can set passwords.
- Users need to show ID to obtain a password.
- ID cards need to be the latest version.

Figure E-23 Busy slide

➣ Practicing for Your Delivery

Surveys indicate that public speaking is most people's greatest fear. However, fear or nervousness can be a positive factor. It can channel your energy into doing a good job. Remember that an audience will rarely perceive that you are nervous unless you fidget or your voice cracks. They are present to hear the content of your talk, so think of the audience, not how you feel.

The presentations you give for the cases in this textbook will be in a classroom setting with 20 to 40 students. Ask yourself this question: Would I be afraid to talk to just one or two of my classmates? Think of your presentation as an extended conversation with several of your classmates. Let your gaze shift from person to person and make eye contact with them. As your gaze drifts around the room, say to yourself, "I'm speaking to one person." As you become more experienced in speaking before groups, you will be able to let your gaze move naturally from one audience member to another.

Tips for Practicing Your Delivery

Giving an effective presentation is not reading a report to an audience. Rather, it requires that you have your message rehearsed well enough so you can present it naturally, confidently, and in tandem with well-chosen visual aids. Make sure that you allow sufficient time to practice your delivery.

- Practice your presentation several times and use your visual aids when you practice.
- Show visual aids at the right time and only at the right time. A visual aid should not be shown too soon or too late. In your speaker's notes, you might even have cues for when to show each visual aid.
- Maintain eye and voice contact with the audience when using the visual aid. Don't look at the screen or turn your back on the audience.
- Use your visual aids and refer to them both in your talk and with hand gestures. Don't ignore your own visual aid.
- Keep in mind that your visual aids should support your presentation, not *be* the presentation. In other words, don't have everything you are going to say on each slide. Use visual aids to illustrate the key points and statistics and fill in with your talk.
- Time check: Are you within time limits?
- Using numbers effectively: Use round numbers when speaking or you'll sound like a computer. Also, make numbers as meaningful as possible: For example, instead of saying "in 84.7 percent of cases," say, "in five out of six cases."
- Don't "reach" to interpret the output of statistical modeling. For example, suppose that you have input many variables into an Excel model. You might be able to point out a trend, but you might not be able to say with certainty that if management employs the inputs in the same combination that you used them, they will get exactly the same results.
- Record yourself, if possible, and then evaluate yourself. If that is not possible, have a friend listen to you and evaluate your style. Are you speaking down to your audience? Is your voice unnaturally high-pitched from fear? Are you speaking clearly and distinctly? Is your voice free of distractions, such as "um" and "you know," "uh, so," and "well"?

- If you use a pointer, either a laser pointer or a wand, use it with care. Make sure that you don't accidentally point a laser pointer in someone's face—you'll temporarily blind them. If you're using a wand, don't swing it around or play with it.

Handling Questions

Fielding questions from an audience can be an unpredictable experience because you can't anticipate all the questions that might be asked. When answering questions from an audience, *always treat everyone with courtesy and respect*, no matter what. Use the following strategies to handle questions:

- Anticipate questions. You can gather much of the information that you need as you draft your presentation. Also, if you have a slide that illustrates a key point but doesn't quite fit in your talk, save it—someone might have a question that the slide will answer.
- Mention at the beginning of the talk that you will take questions at the end of your talk. This will (you hope) prevent people from interrupting your presentation. If someone tries to interrupt you, smile and say that you'll be happy to answer all questions when you're finished or that the next graphic will answer their question. (If, however, the person doing the interrupting is the CEO of your company, you want to stop your presentation and answer the question on the spot.)
- When answering a question, first repeat the question if you have *any* doubt that the entire audience might not have heard it. Then deliver the answer to the whole audience, not just the one person who asked the question.
- Be informative and not persuasive, i.e., use facts to answer questions. For instance, if someone asks your opinion about some outcome, you might show an Excel slide that displays the Solver's output, and then you can use that data as the basis for answering the question.
- If you don't know the answer to a question, don't try to fake it. For instance, suppose someone asks you a question about the Scenario Manager that you just can't answer. Be honest. Say, "That is an excellent question but, unfortunately, it's not one that I'm able to answer." At that point, you might ask your instructor whether he or she can answer the question. In a professional setting, you might say that you'll research the answer and e-mail the answer to the person who asked the question.
- Signal when you are finished. You might say, "I have time for one more question." Wrap up the talk yourself.

Handling a "Problem" Audience

A "problem" audience or a heckler is every presenter's nightmare. Fortunately, such experiences are rare. If someone is rude to you or challenges you in a hostile manner, keep cool, be professional, and rely on facts. Know that the rest of the audience sympathizes with your plight and admires your self-control.

The problem that you will most likely encounter is a question from an audience member who lacks technical expertise. For instance, suppose that you explained how to input data into an Access form, but someone didn't understand the explanation that you gave. In such an instance, ask the questioner what part of the explanation is confusing. If you can answer the question briefly, do so. If your answer to the questioner begins to turn into a time-consuming dialogue, offer to give the person one-on-one input later.

Another common problem is someone who asks you a question that you've already answered. The best solution is to answer the question as briefly as possible and use different words (just in case it's the way in which you explained something that confused the person). If the person persists in asking questions that have very obvious answers, either the person is clueless or is trying to heckle you. In that case, you might ask the audience, "Who in the audience would like to answer that question?" The person asking the question will get the hint.

PRESENTATION TOOLKIT

You can use these forms for preparation, self-analysis, and evaluation of your classmates' presentations (Figures E-24, E-25, and E-26).

Preparation Checklist

Facilities and Equipment

- ❑ The room contains the equipment that I need.
- ❑ The equipment works and I've tested it with my visual aids.
- ❑ Outlets and electrical cords are available and sufficient.
- ❑ All the chairs are aligned so that everyone can see me and hear me.
- ❑ Everyone will be able to see my visual aids.
- ❑ The lights can be dimmed when/if needed.
- ❑ Sufficient light will be available so I can read my notes when the lights are dimmed.

Presentation Materials

- ❑ My notes are available, and I can read them while standing up.
- ❑ My visual aids are assembled in the order that I'll use them.
- ❑ A laser pointer or a wand will be available if needed.

Self

- ❑ I've practiced my delivery.
- ❑ I am comfortable with my presentation and visual aids.
- ❑ I am prepared to answer questions.
- ❑ I can dress appropriate to the situation.

Figure E-24 Preparation Checklist

Evaluating Access Presentations

Course: ________ **Speaker:** ________________________ **Date:** _______

Rate the presentaton by these criteria:
4=Outstanding 3=Good 2=Adequate 1=Needs Improvement
N/A=Not Applicable

Content

_____ The presentation contained a brief and effective introduction.
_____ Main ideas were easy to follow and understand.
_____ Explanation of database design was clear and logical.
_____ Explanation of using the form was easy to understand.
_____ Explanation of running the queries and their output was clear.
_____ Explanation of the report was clear, logical, and useful.
_____ Additional recommendations for database use were helpful.
_____ Visuals were appropriate for the audience and the task.
_____ Visuals were understandable, visible, and correct.
_____ The conclusion was satisfying and gave a sense of closure.

Delivery

_____ Was poised, confident, and in control of the audience
_____ Made eye contact
_____ Spoke clearly, distinctly, and naturally
_____ Avoided using slang and poor grammar
_____ Avoided distracting mannerisms
_____ Employed natural gestures
_____ Used visual aids with ease
_____ Was courteous and professional when answering questions
_____ Did not exceed time limit

Submitted by: ______________________________

Figure E-25 Form for evaluation of Access presentations

Evaluating Excel Presentations

Course: ________ **Speaker:** ________________________ **Date:** _______

Rate the presentaton by these criteria:
4=Outstanding 3=Good 2=Adequate 1=Needs Improvement
N/A=Not Applicable

Content

_____ The presentation contained a brief and effective introduction.

_____ The explanation of assumptions and goals was clear and logical.

_____ The explanation of software output was logically organized.

_____ The explanation of software output was thorough.

_____ Effective transitions linked main ideas.

_____ Solid facts supported final recommendations.

_____ Visuals were appropriate for the audience and the task.

_____ Visuals were understandable, visible, and correct.

_____ The conclusion was satisfying and gave a sense of closure.

Delivery

_____ Was poised, confident, and in control of the audience

_____ Made eye contact

_____ Spoke clearly, distinctly, and naturally

_____ Avoided using slang and poor grammar

_____ Avoided distracting mannerisms

_____ Employed natural gestures

_____ Used visual aids with ease

_____ Was courteous and professional when answering questions

_____ Did not exceed time limit

Submitted by: ______________________________

Figure E-26 Form for evaluation of Excel presentations

Index

D

E

F

G

H

I

J

K

L

M

N

O

P

S

T

U

V

W

Y